THE HUMAN FACTORS OF FRATRICIDE

Human Factors in Defence

Series Editors:
Dr Don Harris, Managing Director of HFI Solutions Ltd, UK
Professor Neville Stanton, Chair in Human Factors at the
University of Southampton, UK
Dr Eduardo Salas, University of Central Florida, USA

Human factors is key to enabling today's armed forces to implement their vision to 'produce battle-winning people and equipment that are fit for the challenge of today, ready for the tasks of tomorrow and capable of building for the future'(source: UK MoD). Modern armed forces fulfil a wider variety of roles thanever before. In addition to defending sovereign territory and prosecuting armed conflicts, military personnel are engaged in homeland defence and in undertaking peacekeeping operations and delivering humanitarian aid right across the world.

This requires top class personnel, trained to the highest standards in the use offirst class equipment. The military has long recognised that good human factors is essential if these aims are to be achieved.

The defence sector is far and away the largest employer of human factors personnel across the globe and is the largest funder of basic and applied research. Much of this research is applicable to a wide audience, not just the military; this series aims to give readers access to some of this high quality work.

Ashgate's *Human Factors in Defence* series comprises of specially commissioned books from internationally recognised experts in the field. They provide in-depth, authoritative accounts of key human factors issues being addressed by the defence industry across the world.

The Human Factors of Fratricide

LAURA A. RAFFERTY
University of Southampton, UK

NEVILLE A. STANTON
University of Southampton, UK

and

GUY H. WALKER
Heriot-Watt University, Edinburgh, UK

CRC Press
Taylor & Francis Group
Boca Raton London New York

CRC Press is an imprint of the
Taylor & Francis Group, an **informa** business

CRC Press
Taylor & Francis Group
6000 Broken Sound Parkway NW, Suite 300
Boca Raton, FL 33487-2742

First issued in paperback 2017

No claim to original U.S. Government works

Version Date: 20160226

ISBN 13: 978-1-138-07583-2 (pbk)
ISBN 13: 978-0-7546-7974-5 (hbk)

Visit the Taylor & Francis Web site at
http://www.taylorandfrancis.com

and the CRC Press Web site at
http://www.crcpress.com

Contents

List of Figures

List of Tables

About the Authors

Dr Laura A. Rafferty
Transportation Research Group, University of Southampton, UK

Laura A. Rafferty completed her undergraduate studies in 2007, graduating with a BSc in Psychology (Hons) from Brunel University, UK. In the course of this degree Laura completed two industrial placements, the second of which was working as a Research Assistant in the Ergonomics Research Group. During this seven-month period Laura helped to design, run and analyse a number of empirical studies being carried out for the Human Factors Integration Defence Technology Centre (HFI DTC) at Brunel. During this time Laura also completed her dissertation, exploring the qualitative and quantitative differences between novices and experts within military command and control. From April 2009 Laura has been employed in the Transportation Research Group, at the University of Southampton, as a Project Assistant for the HFI DTC, working on projects including Naturalistic Decision Making in Teams, Contemporising the Combat Estimate and compiling a Human Factors Methods Database. In July 2011 Laura was awarded a PhD in Human Factors from the University of Southampton, UK.

Professor Neville A. Stanton
Transportation Research Group, University of Southampton, UK

Professor Stanton holds a Chair in Human Factors in the School of Civil Engineering and the Environment at the University of Southampton, UK. He has published over 160 peer-reviewed journal papers and 20 books on Human Factors and Ergonomics. He consults widely on Human Factors and trains Human Factors methods all over the world. Professor Stanton also acts as an expert witness. In 1998, he was awarded the Institution of Electrical Engineers Divisional Premium Award for a co-authored paper on Engineering Psychology and System Safety. The Ergonomics Society awarded him the Otto Edholm medal in 2001, The President's Medal in 2008 and the Sir Frederic Bartlett Medal in 2012 for his contribution to basic and applied ergonomics research. In 2007, The Royal Aeronautical Society awarded him the Hodgson Medal and Bronze Award with colleagues for their work on flight deck safety. Professor Stanton is an Editor of the journal *Ergonomics* and on the editorial boards of *Theoretical Issues in Ergonomics Science*. Professor Stanton is a Fellow and Chartered Occupational Psychologist registered with The British Psychological Society, and a Fellow of The Ergonomics Society. He has a BSc (Hons) in Occupational Psychology from the University of Hull, UK, an MPhil in Applied Psychology and a PhD in Human Factors from Aston University in Birmingham, UK.

Dr Guy H. Walker
School of the Built Environment, Heriot-Watt University

Guy H. Walker is a Lecturer in the School of the Built Environment at Heriot-Watt University, Edinburgh, and his research focuses on human factors issues in infrastructure and transport. He is a recipient, with his colleagues, of the Ergonomics Society's President's Medal for original research. He is also author/co-author of ten books on diverse topics in human factors, including a major text on human factors methods, and is author/co-author of over 60 international peer-reviewed journals.

Acknowledgements

The Human Factors Integration Defence Technology Centre is a consortium of defence companies and universities working in cooperation on a series of defence-related projects. The consortium is led by Aerosystems International and comprises Birmingham University, Cranfield University, the University of Southampton, Lockheed Martin, MBDA, and SEA. The consortium was recently awarded The Ergonomics Society President's Medal for work that has made a significant contribution to original research, the development of methodology, and application of knowledge within the field of Ergonomics.

The research presented in this book originates in part from work conducted by the Human Factors Integration Defence Technology Centre, which was part-funded by the Human Sciences Domain of the UK Ministry of Defence Scientific Research Programme. Any views expressed are those of the author and do not necessarily represent those of the MOD or any other UK government department.

The research presented in this book was also part-funded by the Thomas Gerald Gray Trust partnered with Brunel University – through their award of a bursary for the first 18 months of this research.

We would like to thank all the participants who were involved in the case studies presented. Thanks to all of the staff at the various training institutions described in Chapter 4, Chapter 5 and Chapter 6, as well as numerous Subject Matter Experts at the training institutions and wider military domains, such as Major M. Forster and David Dean.

Special thanks go to Dr Paul M. Salmon, Dr Daniel P. Jenkins and Julie Sewell for feedback on the research presented within this book. Thanks also to Alec James, Katie Plant and Linda Sorensen for taking part in the inter-rater reliability analyses.

List of Abbreviations

1BW	First Battalion Black Watch
1RRF	Royal Regiment of Fusiliers
2RTR	Second Royal Tank Regiment
AB	Air Base
ACO	Allied Command Operations
AGNA	Applied Graph and Network Analysis
Armd	Armoured
ARRSE	The Army Rumour Service
ATC	Air Traffic Controller
ATO	Air Tasking Order
AWACS	Airborne Warning and Control System
Bde	Brigade
BG	Battle Group
BH	Black Hawk
BVR	Beyond Visual Range
C Sqn. QRL	C Squadron Queens Royal Lancers
CAS	Close Air Support
CASEVAC	Casualty Evacuation
CDA	Coordination Demands Analysis
CFAC	Combined Forces Air Component
CID	Combat Identification
CMAQ	Cockpit Management Attitudes Questionnaire
Comms	Communication
Coy	Company
CS	Call sign
CTF	Combined Task Force
CUD	Communication Usage Diagram
Dicker	A single person watching for friendly forces and reporting them to enemy forces
DM	Decision Making
DSA	Distributed Situation Awareness
EAST	Event Analysis of Systemic Teamwork
F3	Famous Five of Fratricide
FEAST	Fratricide Event Analysis of Systemic Teamwork
FSP	Fire Support Position
FST	Fire Support Team
FUP	Forming Up Position
GMLRS	GPS Missile Launcher Rocket System

GPS	Global Positioning System
HESH	High Explosive Squash Head
HFACS	Human Factors Analysis and Classification System
HFI DTC	Human Factors Integration Defence Technology Centre
HMG	Heavy Machine Gun
HQ	Headquarters
hrs	hours
HTA	Hierarchical Task Analysis
ID	Identification
IED	Improvised Explosive Device
IFF	Identification Friend or Foe
IN	Information Network
INCIDER	Integrative Combat Identification Entity Relationship Model
Intel	Intelligence
ISTAR	Intelligence, Surveillance, Target Acquisition, and Reconnaissance
JFC	Joint Fires Cell
JTAC	Joint Tactical Air Controller
Locstat	Location update
MA	Mission Analysis
MCC	Military Command Centre
MoD	Ministry of Defence
NASA	National Aeronautics and Space Administration
NDM	Naturalistic Decision Making
NEC	Networked Enabled Capability
NFZ	No Fly Zone
OC	Operational Commander
OPC	Operation Provide Comfort
PID	Positive Identification
QRL	Queens Royal Lancers
RAF	Royal Air Force
Recce	Reconnaissance
ROE	Rules Of Engagement
ROZ	Restricted Operating Zone
SA	Situation Awareness
SIF	Special Identification Feature
SITREP	Situation Reports
STAMP	Systems Theory Accident Modelling and Process
SME	Subject Matter Expert
SMM	Shared Mental Model
SNA	Social Network Analysis
SOP	Standard Operating Procedure
STAMP	Systems Theory Accident Modelling and Process
STS	Socio Technical System
TAOR	Target Area Of Responsibility

TRADOC	Training and Doctrine Command
UAV	Unmanned Aerial Vehicle
UHF	Ultra High Frequency
USAF	United States Air Force
VID	Visual Identification
WD	Weapons Director
WESTT	Workload, Error, Situational Awareness, Time and Teamwork

Introduction

Introduction to Fratricide

On 25 March 2003, at 00.50, a Challenger II Tank Commander on the outskirts of Basra notices a number of hot spots moving in and out of an object through his thermal imaging equipment. The Commander identifies these as enemy personnel entering and exiting a bunker. He requests and receives permission to engage and a High Explosive Squash Head (HESH) round is fired. Approximately six minutes later the Commander observes a second moving object, which he identifies as an enemy armoured vehicle and a second HESH round is fired destroying the vehicle. (Adapted from the Ministry of Defence 2004)

This is a description of an incident of fratricide in which British fire killed two British military personnel and left a further two seriously injured.

The most commonly utilised definition of 'fratricide' is by the US Army Training And Doctrine Command (TRADOC) Fratricide Action Plan, which defines fratricide as 'the employment of friendly weapons and munitions with the intent to kill the enemy or destroy his equipment or facilities, which results in unforeseen and unintentional death or injury to friendly personnel' (US Army; cited in Wilson, Salas, Priest and Andrews 2007, Kogler 2003, Doton 1996, Hart 2004, Jamieson and Wang 2007, Steinweg 1995, Greitzer and Andrews 2008, 2009). Informally fratricide has been labelled 'friendly fire describing the engagement of troops by ones own side'.

In addition to the injury and death which results from fratricide incidents, researchers have illustrated further negative consequences associated with these incidents, including a loss of operational effectiveness (Greitzer and Andrews 2008, 2009, Kogler 2003, Wilson et al. 2007, Mistry et al. 2009, US Congress 1993) caused by the enforcement of tighter rules of engagement and operating procedures (Wild 1997, Mistry et al. 2009, Hawley, Mares and Marcon 2009); loss of morale in military personnel (Barnet 2009, Wild 1997, Ministry of Defence 2002, Kogler 2003, Hart 2004, Gadsen et al. 2008, US Congress 1993, Jamieson and Wang 2007, Dean and Handley 2006, Greitzer and Andrews 2008, 2009, Mistry et al. 2009); and degradation of relationships with coalition partners (Ministry of Defence 2002, Hart 2004, Barnet 2009, US Congress 1993, Dean and Handley 2006): for example Hart (2004) discusses an incident where Canadian forces considered withdrawing from the conflict in Afghanistan due to the engagement of their troops by American fighter jets.

In recent years the media has played a key role in accentuating public awareness of fratricide incidents (Ministry of Defence 2002, Hart 2004, US Congress 1993, Jamieson and Wang 2007, Dean and Handley 2006, Jarmasz et al. 2009), with headlines such as '2nd "friendly fire" death in 24hrs' (*Sun* 2009) and 'Nine paratroopers shot by British gunship after being mistaken for Taliban' (*Daily Mail* 2008). Unfortunately the headlines reflect the truth, with recent figures stating that 43 per cent of UK deaths in Operation Iraqi Freedom were caused by fratricide (Gadsen et al. 2008). During World War II the number of fratricide-related deaths was thought to be 14 per cent (Steinweg 1995).

Previous research into fratricide has highlighted numerous reasons for the increasing rate of these incidents. There is little to distinguish the enemy from friendly or neutral personnel, the enemy uses equipment that looks very like that of friendly forces and fight in a covert fashion, as opposed to the enemy forming up an opposing front as in traditional Cold War-style warfare (Pirnie et al. 2005, Zobarich, Bruyn-Martin and Lamoureux 2009, Jarmasz et al. 2009, Barnet 2009). Modern warfare is complex (Moffat 2003b, Bar Yam 2003) and technological advances in war-fighting capability means that when friendly or allied troops are incorrectly targeted as enemy, weapons are able to destroy the target from far beyond the human visual identification range (Center for Army Lessons Learned 2005, Ministry of Defence 2002, Kogler 2003, Doton 1996, US Congress 1993, Greitzer and Andrews 2008, 2009, Hawley, Mares and Marcon 2009). According to Barnet, warfare has:

> moved from face-to-face close combat, to lines of musket men, to direct fire at the limits of visual range, to beyond visual range (BVR), where friendly and enemy units are represented iconically on display screens. This distance has complicated combat identification and made it more difficult for soldiers to identify friend from foe. (2009: 313)

The combination of these factors, and the consequences outlined above, emphasize the need to comprehend the causality of fratricide incidents fully and identify preventive measures to protect personnel from such incidents.

There has been a large amount of research undertaken within this domain into the development of technological solutions such as combat identification aids to decipher enemy and friendly personnel (Jamieson and Wang 2007, Wilson et al. 2007, Kogler 2003, Barnet 2009, Pharaon 2009, Hawley, Mares and Marcon 2009, Dzindolet, Pierce and Beck 2009, Neyedli et al. 2009, Rice, Clayton and McCarley 2009). Research, for instance by Kogler (2003), has illustrated the benefit associated with such solutions. In a comparison of technological decision aids, Kogler found that the absence of any decision aid was correlated with the greatest number of fratricide engagements and that a decision aid 'performing at 100 per cent reliability completely eliminated fratricide in the context of this study' (2003: 35). There has also been a great deal of research into the problems associated with technological solutions as, in reality, equipment reliability is less than perfect (Kogler 2003, Parasuraman and Riley 1997, Doton 1996, Greitzer and Andrews 2008, Barnet 2009, Hawley, Mares and Marcon 2009, Dzindolet, Pierce and Beck 2009, Neyedli, Wang, Jamieson and Hollands 2009, Rice, Clayton and McCarley 2009).

In addition to the complexities related to technological solutions, recent research into fratricide has highlighted the need to explore the problem from a Human Factors perspective, focusing on the humans involved and how they interact with the technology (Hart 2004, Wilson et al. 2007, Gadsen et al. 2008, Gadsen and Outteridge 2006, US Congress 1993, Greitzer and Andrews 2008, 2009). For example, Hart argues that:

> the human element, present at every level of decision making before a weapon
> is launched at a target, is the most critical link in the fratricide chain. (2004: 14)

The consequential emphasis on the human negates the ability of a technological 'solution' to fratricide (Hart 2004, Gadsen and Outteridge 2006, Wilson et al. 2007, US Congress 1993, Jamieson and Wang 2007, Doton 1996, Hawley, Mares and Marcon 2009, Rice, Clayton and McCarley 2009). As Doton states:

> believing that the application of technology alone will solve the problem is
> fallacious and foolhardy. (1996: 7)

In light of this it is believed that research is required into the decision-making process and the associated contextual factors involved in fratricide incidents.

Purpose of the Research

This research aims to explore fratricide systemically, looking at not only the core factors involved in these incidents but also the manner in which these factors interact with one another to cause an incident of fratricide. This approach may help to promote a better understanding of interactions within complex systems and help in the formulation of hypotheses and predictions concerning errors in teamwork, particularly incidents of fratricide. This research aims to explore the problem of fratricide at a number of systemic levels, beginning with a small-scale team and progressing to explore a larger team composed of teams. The exploration of multiple systemic levels is widely advocated in the Human Factors domain (Reason 1990, Svedung and Rasmussen 2002, Leveson 2001, 2002, Hollnagel 2005, Von Bertalanffy 1950). The overarching aim of this research is to provide a greater understanding of fratricide incidents in order to lower the rate of occurrence of incidents of fratricide.

Structure of this Book

This book presents a description of the research undertaken into the Human Factors issues associated with fratricide and the resultant conclusions. This Introduction, which provides a summary of the problem of fratricide and explains the purpose and objectives of the research, is followed by eight chapters. An overview of each of the chapters is presented below.

Chapter 1 presents an exploration of fratricide, critiquing current approaches to the study of the problem. A review of the literature from the domains of fratricide, teamwork, Situation Awareness (SA) and schemata is discussed and a fusion of core concepts from these domains enables the development of a prototypical model of fratricide causality.

Chapter 2 presents the application of the model of fratricide causality to an incident of fratricide and an ideal version of events in which the incident of fratricide could not have occurred. The model is able to identify clear divergence between the fratricide and non-fratricide examples, isolating clear performance indicators within the incident of fratricide. The chapter provides initial evidence of the applicability of the model to explaining fratricide causality.

Chapter 3 presents a summarised review of methods currently utilised within the fratricide and wider safety domains. The Event Analysis of Systemic Teamwork (EAST) method, which provides a systemic evaluation of events involving teams, is identified as most appropriate to the exploration of fratricide incidents from the theoretical perspective of this research. Theoretical criteria drawn from the literature are used to illustrate the method's ability to convey the underlying causality of fratricide and adhere to the theoretical stance of this research.

In Chapter 4, EAST is applied to an example of fratricide within a training environment for British Army tank crews. A battle group was observed undertaking numerous pre-deployment training scenarios. In one scenario an incident of fratricide occurred and the EAST methodology was used to compare the tank crew involved with a tank crew who effectively completed the scenario without engaging friendly personnel. This case study represents a small, three-man team illustrative of a low level within the military organisation. Core differences between the effective performance and the fratricide incident were drawn from the case study and the results of the analysis provide further validation for the ability of the prototypical model to explore the complex interactions associated with incidents of fratricide.

A further case study is presented in Chapter 5. This discusses the observation of the Royal Air Force and the British Army undertaking joint pre-deployment training. The scenarios observed involved the two forces working together on Close Air Support missions, thus enabling an exploration of fratricide at a team of teams level. Effective and fratricide performance are compared, again highlighting clear points of divergence within the performances. The chapter highlights the ability of EAST to explore fratricide within a large multi-force situation and the ability of the model to explain the variance between effective performance and the occurrence of an incident of fratricide.

Chapter 6 presents the results of another case study exploring the Royal Air Force and the British Army, in which they undertook training for Close Air Support missions. The focus of this case study was at the Fire Support Team level – a systemic level which sits between the 'small' tank crew study and the large team of teams level discussed previously. The research compared the performances of two teams in a training scenario: one team successfully completed the mission and the second team tasked an Apache attack helicopter to target the team's own location. The causal factors that led to the occurrence of this engagement and the absence of these factors within the team that successfully completed the mission are explored. The results reveal continuity of causality with the results of Chapter 5 and provide further validation of the utility of the EAST method in the investigation of incidents of fratricide.

In Chapter 7, the models developed from the three case studies are compared and contrasted in order to draw out a series of high-level conclusions. The research highlights the commonalities and variances identified between the models and allows for a number of conclusions to be drawn regarding the Human Factors issues underlying incidents of fratricide.

Chapter 9, the Conclusion, provides a summary of the research and the theoretical notions derived from the work. The question of whether the research succeeded in meeting the initial research objectives is discussed alongside the articulation of potential future research paths.

Chapter 1

Fratricide, Expectations, Situation Awareness and Teamwork

Introduction

The purpose of this chapter is to propose foundations for a theory of fratricide causality based on a review of the existing literature. Core suppositions are drawn from current research into fratricide, each of which is explored in relation to prominent work within the wider Human Factors literature. Literature is drawn from multiple sources, including academic journals and texts, government and research reports, and conference papers. Insights gained from wider academia are utilised to develop a model of fratricide causality, and the model's relationship to previous fratricide research is discussed.

Fratricide from a Human Factors Perspective

Recent research by Kogler (2003), Jamieson and Wang (2007), Dean and Handley (2006), Gadsen and Outteridge (2006), US Congress (1993) Masys (2006) and Wilson et al. (2007) has highlighted the importance of exploring fratricide from a systems perspective hypothesising that multiple causal factors interact in non-linear ways to cause an incident of fratricide.

The Importance of Expectations

Although this book focuses on the systemic evaluation of fratricide, research into the individual decision-making processes involved in fratricide has identified a number of interesting conclusions. Famewo, Matthews and Lamoureux (2007) and Dean and Handley (2006) provide in-depth reviews of the individual decision-making process involved in shoot, no-shoot decisions. Both researchers draw out the importance of expectations and confirmatory bias in the decision-making process (Famewo, Matthews and Lamoureux 2007, Dean and Handley 2006), arguing that expectation impacts on the way in which individuals interpret the situation they are in, and the way they attend to and integrate information. Dean and Handley (2006) present the Integrative Combat Identification Entity Relationship Model (INCIDER) model depicting the human decision-maker involved in incidents of fratricide. The model states that the decision-maker will go through a:

comprehension process by which the decision maker compares his expectation
with the sensory input (i.e. compares his previous SA with new sensory input to
derive a revised SA). (Dean and Handley 2006: 17)

The INCIDER model also represents myriad impacts upon the decision-maker,
ranging from personality factors to weather conditions, thereby emphasising the
complexity of the problem of fratricide (Dean and Handley 2006).

Famewo, Matthews and Lamoureux (2007) focus on the manner in which
decision-makers combine and assign importance to information, claiming that
a core aspect of fratricide prevention is ensuring that information is correctly
received and integrated. The importance attached to appropriate information
integration is replicated across the fratricide domain (Bolstad, Endsley and Cuevas
2009). Famewo, Matthews and Lamoureux suggest that 'not all information is of
equal value' and propose that people use their previous experience to determine
the value of information (2007: 30). Building upon the role of expectations, the
research explores the confirmatory bias and its impact upon incidents of fratricide.
Famewo, Matthews and Lamoureux define confirmatory bias:

> [W]hen people have a strong opinion or belief about the state of the world,
> they are more likely to seek evidence or cues that confirm this belief. Evidence
> in contrast to their opinion is perceived as an outlier that should be ignored.
> (2007: 33)

For an in-depth review of the confirmatory bias and its impact upon fratricide the
reader is also referred to Famewo, Matthews and Lamoureux (2007).

Greitzer and Andrews (2009) continue the emphasis on the importance of
expectations to fratricide incidents, suggesting that the impact of such expectations
is heightened in situations of acute stress. Their research discusses numerous
studies which have shown that stress can lead to tunnel vision, causing individuals
to focus their attention on certain cues to the detriment of other cues. Greitzer and
Andrews (2009) argue that expectancy can cause:

> selective perception as well as biased decisions or responses to situations in the
> form of other cognitive biases like confirmation bias (the tendency to search
> for or interpret information in a way that confirms one's preconceptions) or
> irrational escalation (the tendency to make irrational decisions based upon
> rational decisions in the past). (Greitzer and Andrews 2009: 180)

Greitzer and Andrews suggest that the importance of such cognitive biases is
undeniable, since these cognitive biases can lead soldiers to 'start looking for
reasons to fire instead of reasons not fire' (2009: 183), suggesting that in-depth
exploration is needed to understand the impact of expectations on fratricide.

Teamwork

Research into the individual processes involved in fratricide has provided a number of interesting insights. However, the focus of this research is on the systemic aspects of fratricide in line with current academic as well as government guidance, for example the US Congress (1993) emphasised the need to view the problem as an extended process situated across systemic levels. According to Greitzer and Andrews 'combat ID and maintaining situation awareness (SA) in general, are highly distributed, collective tasks' (2009: 229).

Research by Wilson et al. (2007) explores the wider issues associated with fratricide incidents, developing a model of fratricide representing the military system as a whole including organisational, technological and task factors. The focus of the model developed by Wilson et al. (2007) is on the teamwork aspects associated with fratricide, but the model also illustrates the roles of other organisational factors in these incidents. Whereas the previous discussions (Dean and Handley 2006, Famewo, Matthews and Lamoureux 2007) focused on how information was utilised in decision-making, Wilson and his colleagues also explore in their research how that information is gained and the team processes involved in information transfer. Their model of fratricide causation explains fratricide with respect to breakdowns in shared cognition; specifically the teamwork processes of communication, cooperation and coordination.

The taxonomy used by Wilson et al. (2007) has now been extended to identify the possible ways in which these factors could break down. Rather than simply ask team members whether they engage in closed loop communication, the framework is extended to ask the question: If not, why not? Possible reasons for breakdowns in communication are presented in order to guide the analyst in exploring the wider causality of the breakdown (Wilson, Salas and Andrews 2009). The framework used by Wilson et al. has also been extended beyond an error identification tool to provide possible training strategies to mitigate teamwork breakdowns. Wilson, Salas and Andrews (2009) have mapped the core breakdowns associated with communication, coordination and cooperation onto prominent training strategies from the Human Factors domain. In this manner Wilson, Salas and Andrews (2009) have provided a methodology to identify specific problems in teams involved in fratricide and identify appropriate training mechanisms which focus on these breakdowns.

The value of the exploration of fratricide carried out by Wilson et al. (2007) is highlighted in the research of Zobarich, Lamoureux and Bruyn-Martin (2007). Zobarich and his colleagues utilised the taxonomy devised by Wilson et al. to identify appropriate behavioural markers that should be measured and conducted a task analysis to identify when these behaviours should be measured. The research developed a method to measure team performance based upon these behavioural markers, together with a five-point scale to rate each marker during team performance. The research of Wilson et al. and Zobarich et al. provides a clear foundation for measuring the factors involved in fratricide. However, current research has emphasised the need to explore the interactions between these causal factors. This questions the utility of taxonomic approaches.

Interactions

Masys (2006) provides an interesting exploration of the problem of fratricide, and wider accident aetiology, focusing on the non-linear interactions among components in complex systems. He approaches fratricide utilising a dual perspective of actor–network theory and complexity theory to explore the social and technical aspects involved in such incidents and the interactions and relationships between these (Masys 2006). Masys (2006) argues that fratricide analysis should not focus on human error proposing that fratricide occurs in a complex system:

> viewed as a network construct of heterogeneous elements relationally interconnected via aligned and opposing interests. (Masys 2006: 377)

In light of this, exploration of fratricide should focus on the emergence and relationships within the military system. Masys views these relationships as 'complex interactions such that inputs and outputs are not proportional', and suggests that due to this, research should focus on 'emergence, non-linearity and co-evolution' (Masys 2006: 377).

Research by Gadsen and Outteridge (2006) and Gadsen et al. (2008) continues the emphasis on the interaction between causal factors and the need to explore fratricide from a systemic perspective. Their research argues that the causality behind fratricide incidents consists of multiple errors across multiple levels of the system, rather than the solider firing the weapon alone. Following from this multi-causality argument, their research highlights the need to explore fratricide incidents with respect to the wider system in which they took place and emphasises that 'the relationships between the factors must be addressed' (Gadsen and Outteridge 2006: 14). In addition to this, their research highlights the cognitive processes involved in fratricide and, in particular, the role of expectations (in line with research by Dean and Handley 2006 and Famewo et al. 2007), suggesting that expectations impact on the way in which people search for, perceive and integrate information (Gadsen and Outteridge 2006, Gadsen et al. 2008).

In order to investigate fratricide causality Gadsen et al. (2008) explored a series of case studies of real-life incidents of fratricide, utilising the Fratricide Causal Analysis Scheme. This categorisation consists of 12 groupings: command and control; procedures; equipment/technology; Situation Awareness; misidentification; physical/physiological factors; pre-deployment preparation; teamwork; environmental factors; communications/information; platform configuration; and cognitive factors (Gadsen et al. 2008). The research concluded that breakdowns in the *communications/information* category were the most prevalent causal factor in the case studies of fratricide explored (Gadsen et al. 2008).

Suppositions

From the previous research into fratricide a number of key suppositions can be identified:

1. Fratricide is complex, multi-causal and is the result of problems at multiple levels of the military system.
2. Further research is needed into the interactions between causal factors.
3. Expectations are important to fratricide and to Situation Awareness (SA).
4. Situation Awareness is an important factor of fratricide causality.
5. Teamwork is an important factor of fratricide causality.

From this initial review of the fratricide literature it is clear that fratricide is a problem emerging from within the military system as a whole and is the outcome of numerous interconnected causal factors, including expectations, SA and teamwork. The next part of this chapter provides an exploration of each supposition within the wider literature in an attempt to identify currently available models that may enable an investigation of the complex causality associated with incidents of fratricide.

Wider Literature

Complex, Multi-Causal, Multiple Levels and Interactions

The first two propositions represent the underlying theoretical assumptions of general systems theory. Von Bertalanffy proposed 'a new scientific doctrine of wholeness' (1950: 142) called general systems theory, arguing that systems should be explored based upon the proposition that 'the whole is more than the sum of its parts' (142). General systems theory defines a system as 'a complex of interacting elements' (143) proposing that each of these elements will behave differently when explored in isolation from how it would behave within the system as a whole. Each element's behaviour is dependent upon the interaction of all elements within the system. Von Bertalanffy asserts that studying elements in isolation is insufficient:

> You cannot sum up the behaviour of the whole from the isolated parts, and you have to take into account the relations between the various subordinated systems and the systems that are subordinated to them in order to understand the behaviour of the parts. (Von Bertalanffy 1950: 148)

Heylighen and Joslyn (1992) describe general systems theory as a theory which argues against reductionism and emphasises holism, positing that elements within a system are constantly interacting, evolving and producing emergent properties. It has been proposed that accidents can be viewed as an aberrant emergent property arising as a result of constant interaction and evolution within such systems (Qureshi 2007).

The systems perspective is becoming an increasingly popular theoretical stance supported by researchers such as Hollnagel (1993) and Hancock (1997) and it is being applied to numerous domains including SA (Stanton, Stewart et al. 2006, Salmon, Stanton, Walker, Baber, Jenkins, McMaster and Young 2008); Situation Awareness and anaesthesia (Fioratou et al. 2010); Command and Control interface design (Jenkins et al. 2008) and human error identification (Stanton and Baber 1996).

The systems approach is purported to be particularly fitting to the exploration of the fratricide incidents, as is discussed in the fratricide literature (see above), and to the wider military system within which fratricide occurs, as emphasised by Bar Yam (2003) and Jensen (2003). Bar Yam defines modern warfare as 'a complex encounter between complex systems in complex environments' (2003: 1), hypothesising that in order to understand modern warfare one must accept that it is not a simple linear construct, but a complex system defined as:

> multiple interacting elements whose collective actions are difficult to infer from those of the individual parts, predictability is severely limited, and response to external forces does not scale linearly with the applied force. (Bar Yam 2003: 1)

In order to explore the causality associated with fratricide incidents, this book presents a system oriented, rather than an individual-oriented, analysis of the problem.

Expectations

The third supposition raised by the fratricide literature is the role played by expectations in the causality of incidents of fratricide. The importance of expectations to safety is prominent throughout the wider safety and error literature (Woods et al. 1994, Mitchell and Flin 2007). A model that is frequently utilised to explore and explain the role of expectations is Neisser's Perceptual Cycle model (Stanton, Chambers and Piggot 2001, Jenkins, Salmon et al. 2011). The model has previously been applied to accidents and error within other domains such as aviation (Banbury, Dudfield and Lodge 2002), and to military decision-making within a planning environment (Freeman and Cohen 1994).

Neisser's Perceptual Cycle model   Neisser's Perceptual Cycle model argues that the world contains more information than anyone can process at once. The information that an individual chooses to attend to is guided by 'schemata', unique structures which focus an individual's attention on 'certain aspects of the environment rather than others, or indeed to notice anything at all' (Neisser 1976: 9). Schemata are unique to each situation, they provide a generic template for action, but the exact action taken is determined by the specific interaction with the environment at that moment (Neisser 1976). These schemata are also unique to each individual as they are:

> developed by experience; everyone's experiences are different; therefore we must all be very different from one another. Since every persons perceptual history is unique we should all have unique cognitive structures. (Neisser 1976: 187)

Due to this specific nature of schemata the information an individual chooses to attend to can vary drastically. Neisser hypothesises that every object can hold a range of affordances and the dominant affordance seen is dependent upon who is viewing the object. The meaning assigned to an object (or piece of information) depends upon the schemata activated (Neisser 1976). Neisser discusses the way in which two people can view the same situation in very different ways such as how a chess master sees the correct move to make on a chess board where as a baby just sees something is in front of him:

> The differences among these perceivers are not matters of truth and error but of noticing more rather than less. The information that specifies the proper move is as available in the light sampled by the baby as by the chess master, but only the master is equipped to pick it up. (Neisser 1976: 180–81)

Once an individual has interacted with the world, guided by a schema, the information gained then serves to update that schema, 'people act on what they know and are changed by the consequences of their actions' (Neisser 1976: 3). In summary, action and perception 'depend upon pre-existing structures, here called schemata, which direct perceptual activity and are modified as it occurs' (Neisser 1976: 14).

Norman's (1981) interpretation of Neisser's Perceptual Cycle model   Norman (1981) applied Neisser's (1976) model to error, hypothesising that schemata are triggered when the environment matches those schemata sufficiently, stating that a perfect match is clearly not required or there would never be inappropriate triggering of schemata. Norman discusses the manner in which inappropriate schemata activation can occur, dividing these into three core types:

1. Inappropriate activation: when 'an action entirely appropriate for a situation is being performed, except that this is not the current situation' (Norman 1981: 7) due to an inappropriate classification of the situation.
2. Unintentional activation: when 'a schema may be unintentionally activated, thereby causing an action to intrude where it is not expected' (Norman 1981: 7).
3. Inappropriate timing: when 'a schema may be properly selected and activated but lead to a slip because it is triggered improperly, either at the wrong time or not at all' (Norman 1981: 10).

The research of Norman and Neisser would present fratricide causality as a co-evolutionary process, presenting the world and the individual in a tightly coupled evolution. This perspective fits with the individual decision-making research of Dean and Handley (2006) and Famewo et al. (2007) presented earlier, and although the theory will have to be modified to represent the need to look at the system as whole (not just a single individual decision-maker) the theory also sits well with the importance placed on SA within the fratricide literature as described in the next part of this chapter.

Situation Awareness

The importance of SA to safety has been highlighted within the fratricide literature (Greitzer and Andrews 2009, Bolstad, Endsley and Cuevas 2009, Barnet 2009) and is common throughout the wider literature as well (Salmon, Stanton, Walker and Jenkins 2009, Stanton, Chambers and Piggott 2001). The most frequently referenced definition of SA is that of Endsley who defines SA as:

> the perception of the elements in the environment within a volume of time and space, the comprehension of their meaning and the projection of their status in the near future. (Endsley 1995: 36)

This model is based on an understanding of an individual's SA, but it has been utilised to explore team SA, defined as:

> the degree to which every team member possesses the SA required for his or her responsibilities. (Endsley 1995: 31)

There are a number of problems associated with this definition of team SA: for example, the scaling up of an individual model of SA to explore team SA will still represent an individual orientation (Salmon, Stanton, Walker and Jenkins 2009) that may be unable to explore the specific processes involved in SA at the team level (Salas, Muniz and Prince 2006). Salmon et al. (2009) undertook an in-depth review of SA models applicable to complex systems. The research concluded that current models of team SA, such as Endsley's, were unable to account for the process, content and factors impacting SA development in teams, focusing on SA within individual team members' heads rather than viewing SA as a phenomenon at the team level. Salmon, Stanton, Walker and Jenkins (2009) argued that distributed models of SA, in which SA is seen as team-level property distributed across team members, were more suitable.

Stanton, Stewart et al. (2006) put forward a theory of Distributed Situation Awareness (DSA), which states that SA is an emergent property arising from the interaction and coordination of agents within a system. The theory suggests that analysis should not, therefore, explore SA in a system: rather, it should explore the interaction of SA within the system under analysis (Stanton, Salmon, Walker and Jenkins 2009a). The theory of DSA holds that SA is distributed across a system and is 'a dynamic and collaborative process binding agents on tasks on a moment-by-moment basis' (Stanton, Stewart et al. 2006: 1). From the DSA perspective, agents within a system hold compatible rather than shared SA. That is, due to the diversity of roles and skills within a system, not all agents need to hold the same SA (Stanton, Stewart et al. 2006).

The main distinction between DSA and shared SA is the different purposes of agents within the system; shared SA posits that agents need the same information, since they are working for the same purpose, whereas DSA recognises that within different systems, SA and goals are varied and diverse but may for some periods of a task be compatible (Stanton, Salmon, Walker and Jenkins 2009a, 2009b). In addition to the varied goals of agents within a system, the theory argues that individuals interpret information in different ways due to their individual schemata, which provides further evidence against the notion of shared SA (Stanton, Salmon, Walker and Jenkins 2009b). Endsley's model of SA has been successfully used to explore team SA. Within this research (and in alignment with Salmon, Stanton, Walker and Jenkins 2009), however, the complexity of the problem, and the systemic perspective adopted emphasise the relevance of system-oriented approaches like that of Stanton, Stewart et al. (2006).

Teamwork

Teamwork is Important to Safety

The systemic perspective taken by this research and previous research into the problem of fratricide (Wilson et al. 2007, Famewo et al. 2007, Greitzer and Andrews 2009) emphasises the need to explore the teamwork issues involved in fratricide. In line with the Wilson et al. (2007) model, and the research by Famewo et al. (2007), this book views fratricide as a systems problem but has chosen to focus upon the teamwork aspects of fratricide as the main unit of analysis. Teamwork is hypothesised to be 'a core component of team effectiveness because it compliments taskwork in the achievement of tactical and strategic objectives' (Stagl et al. 2006: 117). It has been argued that non-technical skills such as teamwork are 'responsible for maintaining safety' (Flin, Yule et al. 2006: 145) and therefore may play a role in preventing incidents such as fratricide. Crichton and Flin define non-technical skills as 'social, cognitive skills that are crucial to safe and effective management by teams in emergency situations' (2004: 1317). Non-technical skills include factors such as situation awareness, decision-making, communication and leadership (Crichton and Flin 2004). The importance of these non-technical skills to safety has been illustrated across numerous domains including operating theatres (Flin, Yule et al. 2006), aviation (Flin and Maran 2004) and the nuclear power domain (O'Connor et al. 2008), further emphasising the utility of exploring the role of such non-technical, teamwork skills in fratricide incidents.

Teamwork Revolves around Core Processes

Salas and colleagues present an extensive and in-depth exploration of teams and their performance. In a decisive review of teamwork Salas, Sims and Burke (2005) derive the 'Big 5 of Teamwork', arguing that five factors and three core coordinating mechanisms are essential to teamwork. According to Salas and his colleagues, leadership, mutual performance monitoring, back-up behaviour, adaptability and team orientation are the five most important factors to effective team performance. They argue that shared mental models, closed loop communication and mutual trust are also needed within a team to ensure that the 'Big 5' factors exist and are effectively undertaken. The notion of shared mental models fits with the research presented so far within this book in that the importance of an internal cognitive construct is recognised. However, the 'shared' component does not align with the 'distributed' theories of cognition discussed in the earlier parts of this chapter. The identification of a range of core components in teamwork is an interesting concept and the following section attempts to derive a series of components which better fit with the theoretical premises underlying this book.

Core Component Review

In addition to the exploration of the prominent work by Salas and colleagues a wider review of the teamwork literature was undertaken. Utilising the grounded theory approach established by Glaser and Strauss (1967), over 80 pieces of contemporary literature were classified into core teamwork categories. Developing theory using the grounded theory perspective involves the use of open coding, defined as:

> the analytic process through which concepts are identified and their properties and dimensions are discovered in data. (Strauss and Corbin 1998: 101)

The literature review and categorisation process consisted of six core phases, in line with Glaser and Strauss's (1967) open coding methodology:

1. Search for literature relating to teamwork.
2. Identify the core focus(es) of the text.
3. Label the reference with the core factors it discusses.
4. Continue for each piece of literature.
5. Review series of identified factors.
6. Merge factors into higher-level categories based upon underlying dimensions.

This review identified 40 high-level categories representing different factors that may impact on effective team performance. Each individual category was given a numerical value based upon its frequency of occurrence within the literature reviewed. The results of this analysis are presented in Figure 1.1 below, which illustrates each of the high-level factors and the proportion of discussion they hold within the literature.

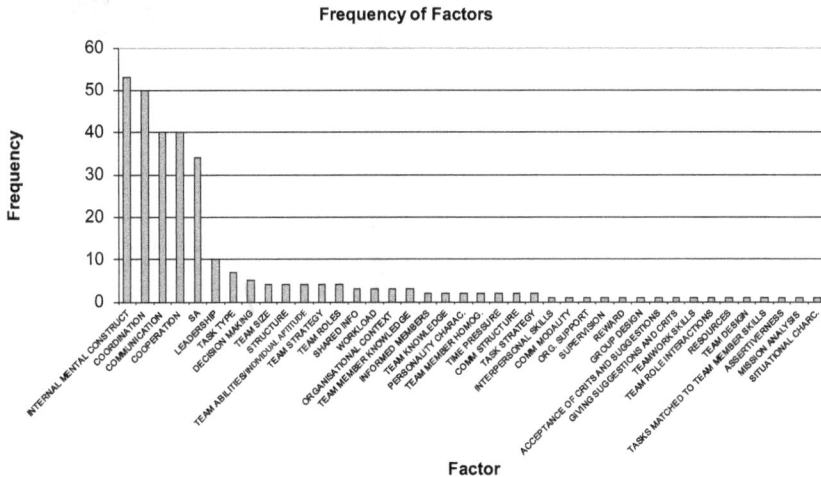

Figure 1.1   Frequency of factors in the literature

The illustration reveals that five factors appear to account for a high level of the discussion present in the Human Factors literature. The aim of this review is not to identify all of the factors that may contribute to team performance within fratricide incidents, but rather to identify those factors which account for the majority of impacts on team performance in the majority of cases. According to the literature reviewed, the factors that capture the majority of discussions around teamwork are: Communication, Cooperation, Coordination, Schemata and SA; there are numerous other factors that may impact on the process, but these five appear to have the greatest impact.

Annett and Stanton (2000) review contemporary teamwork approaches in a similar manner to the approach taken here, drawing key factors from existing models. Their review highlighted the importance of communication, SA, schemata and decision-making. Indeed, a number of researchers agree with the importance of at least four of the five factors (see Serfaty, Entin and Volpe 1993, Fischer, McDonnel and Orasanu 2007, Fiore et al. 2003, Salas, Muniz and Prince 2006, Svensson and Andersson 2006).

Five Factors

Each of the factors is defined, in relation to the theoretical stance of this thesis, below in Figure 1.2:

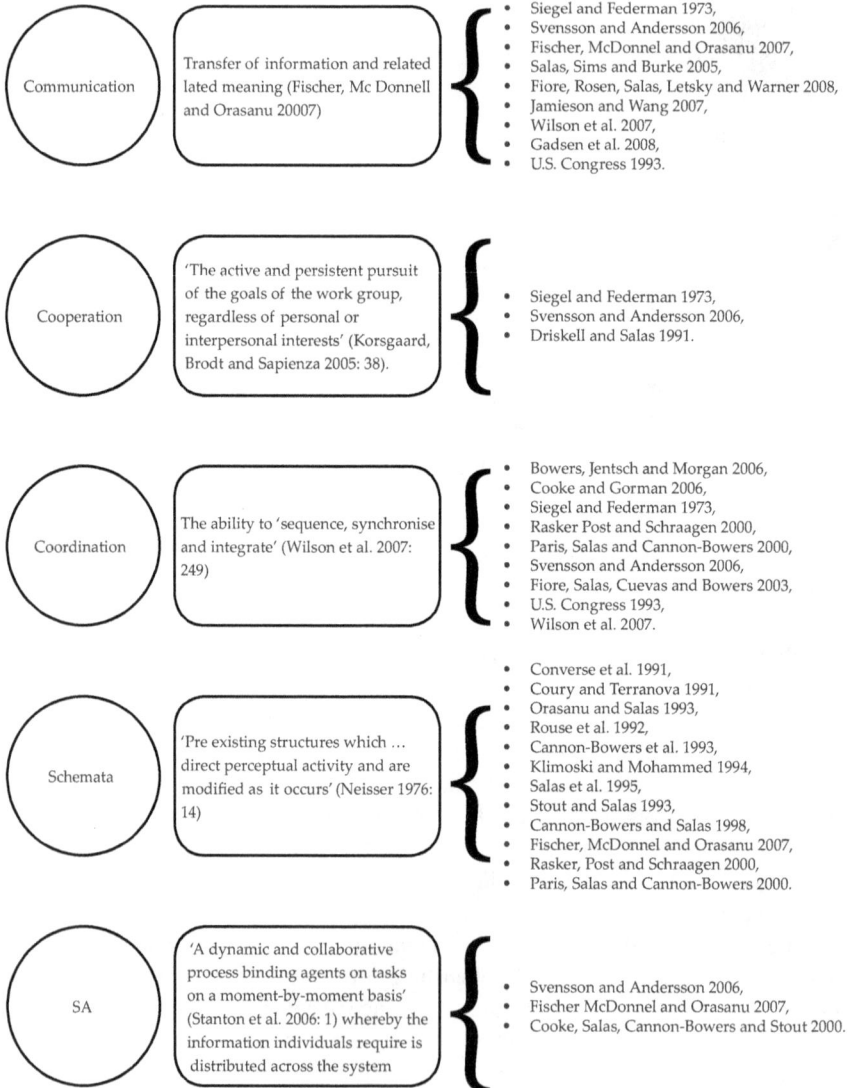

Communication	Transfer of information and related lated meaning (Fischer, Mc Donnell and Orasanu 20007)

- Siegel and Federman 1973,
- Svensson and Andersson 2006,
- Fischer, McDonnel and Orasanu 2007,
- Salas, Sims and Burke 2005,
- Fiore, Rosen, Salas, Letsky and Warner 2008,
- Jamieson and Wang 2007,
- Wilson et al. 2007,
- Gadsen et al. 2008,
- U.S. Congress 1993.

Cooperation	'The active and persistent pursuit of the goals of the work group, regardless of personal or interpersonal interests' (Korsgaard, Brodt and Sapienza 2005: 38).

- Siegel and Federman 1973,
- Svensson and Andersson 2006,
- Driskell and Salas 1991.

Coordination	The ability to 'sequence, synchronise and integrate' (Wilson et al. 2007: 249)

- Bowers, Jentsch and Morgan 2006,
- Cooke and Gorman 2006,
- Siegel and Federman 1973,
- Rasker Post and Schraagen 2000,
- Paris, Salas and Cannon-Bowers 2000,
- Svensson and Andersson 2006,
- Fiore, Salas, Cuevas and Bowers 2003,
- U.S. Congress 1993,
- Wilson et al. 2007.

Schemata	'Pre existing structures which ... direct perceptual activity and are modified as it occurs' (Neisser 1976: 14)

- Converse et al. 1991,
- Coury and Terranova 1991,
- Orasanu and Salas 1993,
- Rouse et al. 1992,
- Cannon-Bowers et al. 1993,
- Klimoski and Mohammed 1994,
- Salas et al. 1995,
- Stout and Salas 1993,
- Cannon-Bowers and Salas 1998,
- Fischer, McDonnel and Orasanu 2007,
- Rasker, Post and Schraagen 2000,
- Paris, Salas and Cannon-Bowers 2000.

SA	'A dynamic and collaborative process binding agents on tasks on a moment-by-moment basis' (Stanton et al. 2006: 1) whereby the information individuals require is distributed across the system

- Svensson and Andersson 2006,
- Fischer McDonnel and Orasanu 2007,
- Cooke, Salas, Cannon-Bowers and Stout 2000.

Figure 1.2   Definition of five core factors

These causal categories were quantifiably shown to be the most frequent factors discussed within the teamwork literature reviewed. Although this does not ensure that they are the most important factors – it simply shows that they are the most cited factors – it is a clear indication of importance and provides an excellent starting ground for the development of a theory of fratricide.

Some researchers may argue that these five factors have only been identified due to circular reasoning. Researchers only mention five factors that define teamwork, so they are the only ones mentioned by other researchers (such as this book) and therefore are the only factors mentioned in future works. However, this argument can be countered by the current research, as 40 separate factors were found within the literature on teamwork, clearly showing that it is not the case that only five factors are ever discussed. It must also be mentioned here that simply because these five factors are most prominent in the literature does not mean that they are the most important. Such an exploration of past literature does, however, undoubtedly provide a solid foundation for identifying the key factors affecting teamwork.

Interactions

The initial investigation into fratricide research emphasised the need to explore the way in which causal factors interact (Gadsen et al. 2008, Gadsen and Outteridge 2006, Masys 2006). Although the framework of Wilson and his colleagues does not present an examination of the manner in which factors interact, their description of the framework does discuss some core interactions between communication, cooperation and coordination. Wilson et al. (2007) argue that adequate information exchange is what enables the development of shared cognition – a shared mental model within the team – and thus allows for a greater level of team SA. Stout, Cannon-Bowers and Salas (1999) echo this sentiment, linking effective communication to effective SA.

The emphasis on interactions is reinforced by the literature surrounding expectations. Neisser's (1976) model of the Perceptual Cycle highlights the importance of the relationships between factors. Factors are not independent – they are dynamic and interdependent, they affect one another and this must be appreciated in a model of teamwork. The importance of interactions is also highlighted in the theory of DSA, which posits that SA is innately tied to schemata, as individuals interpret information in different ways due to their differing goals and tasks: 'individual team members experience a situation in different ways ... defined by their own personal experience, goals, roles, training, knowledge, skills and so on' (Stanton, Salmon, Walker and Jenkins 2009b: 51). The tight coupling of SA and schemata is emphasised in the models of both Neisser (as discussed earlier) and Smith and Hancock (1995) who developed a theory of SA centred on the notion that schemata guide our interaction with, and interpretation of, the world. Endsley (1995) argues that team SA is more reliant on schemata than it is on verbal communication, highlighting the importance of schemata, as well as the relationship between SA and schemata. Later work by Endsley and colleagues has continued this notion, emphasising the importance of SA to fratricide avoidance in particular (Bolstad, Endsley and Cuevas 2009).

General system theory is based on the premise that interactions are highly important; the theory argues that non-linear interactions and evolutions are an integral part of complex systems, with even minor factors being capable of producing considerable changes in the system (Von Bertalanffy 1950). The actual behaviour of such systems becomes difficult to 'model' because they do not adhere to simplistic 'cause and effect' logic. This makes the exploration of the relationships between factors even more important.

Within the teamwork literature there are numerous accounts illustrating the importance of the interaction between team factors. According to Svensson and Andersson (2006) communication is important for effective cooperation and coordination within teams and enables team members to develop and maintain appropriate SA. The link between communication and coordination and SA has been emphasised by other researchers such as Espinosa, Lerch and Kraut (2002) and Patrick, James and Ahmed (2006). Communication and schemata also have an important relationship within teamwork and team decision-making. Flin, Slaven and Stewart (1996) identify communication and schemata as vital in offshore emergency decision-making. In addition to being important to teamwork, coordination within a team is also important to other aspects of team performance such as communication (Fiore et al. 2003, Cooke and Gorman 2006).

The existing literature has been used to identify generic factors that may have caused broken links in fratricide. The 'errors' explored by the model are not generic error or teamwork errors; rather, the errors discussed relate to inadequate relationships between these factors. The discussion of the factors, and the way in which they interact with one another, highlights their cyclical nature. It is the interrelationships between the factors, and the inter-reliance of these links on one another that is of interest within this model.

In order to investigate the interactions between factors, the wider literature was explored once more. Grounded theory was utilised for a second time to categorise the core ways in which factors were proposed to interact with one another in the literature. Data from the literature review enabled the links between these factors to be established and quantitative values to be placed on the links (again from the frequency of discussion of these links within the literature). The factors were developed into a matrix of from and to.

In order to validate the model and further explore the factors, a social network analysis (Driskell and Mullen 2005) was undertaken on the factors and their interactions. Social network analysis is a method used to produce models illustrating the non-linear relationships between factors in a system. Social network analysis aligns with general systems theory in that it describes interactions within the context in which they occur (Driskell and Mullen 2005). The matrix of relationships was inputted into a social network analysis software application called Applied Graph and Network Analysis (AGNA) (Benta 2003). The AGNA software produced a visual representation of the relationships between the factors and the comparative strength of these links. Figure 1.3 below illustrates this interaction:

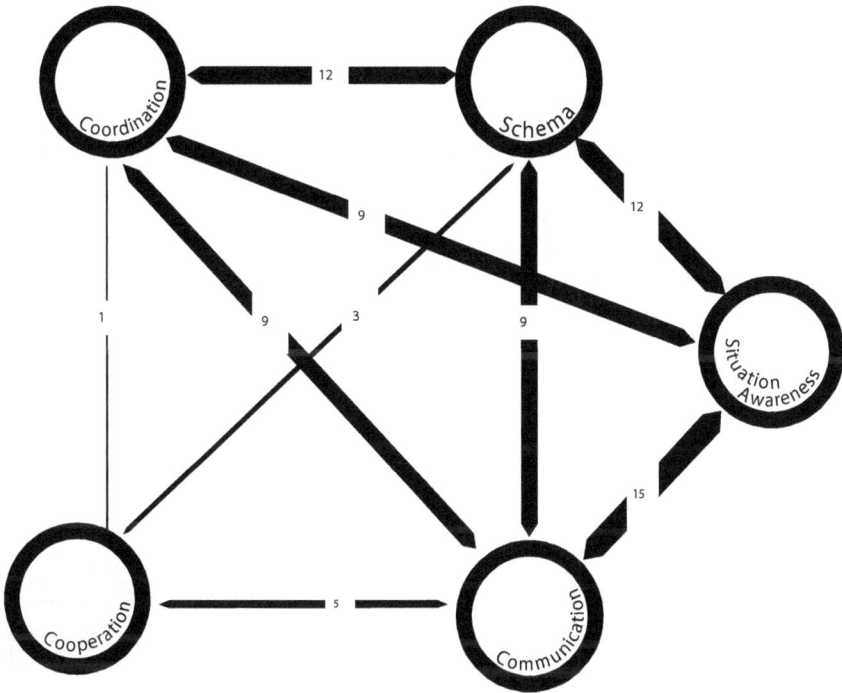

Figure 1.3   Interaction of key factors

The lines on the model represent links between factors explicitly stated within the literature. The numbers beside the lines and the thickness of the lines represent the number of times the link was coded within the literature, that is, the number of separate pieces of research literature that explicitly stated the link between the two factors.

Once the social network had been created, a number of statistics could be derived from it using graph theory (Driskell and Mullen 2005). Within this analysis two metrics were derived:

$$Centrality = \frac{\sum_{i=1 j=1}^{g} \delta ij}{\sum_{j=1}^{g} (\delta ij + \delta ji)}$$

Where g is the number of elements in the network (its size) and δij is the number of edges (e) on the shortest path between elements i and j (Houghton, Baber, McMaster et al. 2006). The centrality metric is utilised to illustrate the prominence of an element within the network, how tightly coupled, or connected, the element is with other elements (Walker, Stanton and Salmon 2011). The centrality metric measures the links between elements, whereas the sociometric status illustrates the strength of these links, how often they are utilised.

$$Sociometric\ status\ \frac{1}{g-1}\sum_{j=1}^{g}(xji+xij)$$

Where g is the total number of elements in the network, i and j are individual elements, and xij are the number of links between element i and j (Houghton et al. 2006).

Figure 1.4 below represents the sociometric status and centrality metrics for the five core factors involved in teamwork:

Social Network Analysis Graph Theory Metrics

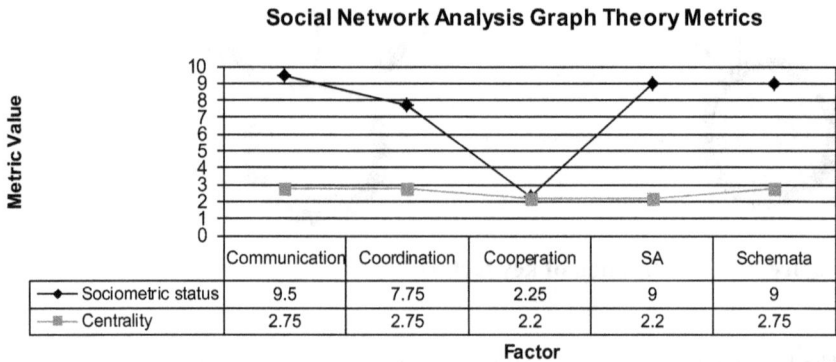

	Communication	Coordination	Cooperation	SA	Schemata
Sociometric status	9.5	7.75	2.25	9	9
Centrality	2.75	2.75	2.2	2.2	2.75

Factor

Figure 1.4   Social network analysis of F3 model

The graph theory metrics illustrate that communication is the most central factor and also the factor with the highest level of sociometric status: it holds the most connections and the most frequently utilised connections with other factors in the network.

From the literature review of the key factors affecting teamwork and the ways in which these factors affect one another, a new model of fratricide was developed. This model, the Famous Five of Fratricide model (F3) can be seen in Figure 1.5:

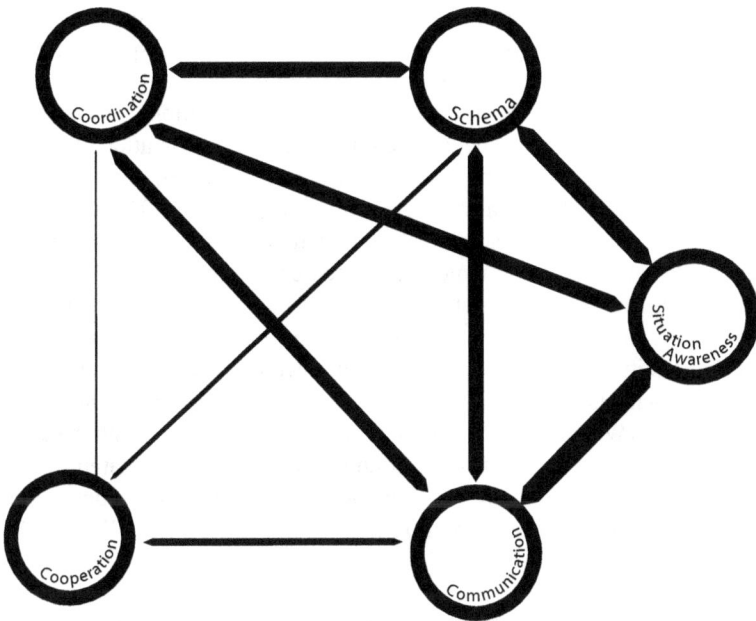

Figure 1.5   Famous Five of Fratricide (F3) model

Relationship of Model to Current Fratricide Literature

Dean and Handley 2006 and Famewo et al. 2007

The importance attached to expectations (schemata) and SA presented within the individual focused fratricide research shows that these represent core components of the F3 model. The relationship between schemata and SA is one of the strongest links (in terms of frequency of occurrence) between factors within the model. The research of Dean and Handley (2006) and Famewo et al. (2007) focuses on the individual, and so the F3 model presented above incorporates and builds upon the relationship between schemata and SA in order to provide a theoretical explanation fitting at the team level.

Wilson et al. 2007

The (2007) model of shared cognition breakdowns by Wilson and colleagues presents a fundamental exploration of core breakdowns which may contribute to the occurrence of incidents of fratricide. Unfortunately this model does not align with the theoretical stance put forward within this book for a number of reasons. Firstly, within this book an emphasis is placed upon general systems theory and the utility of Neisser's Perceptual Cycle as a model to explain the way in which interaction with the world occurs. In light of this theoretical standing the theory of DSA was identified as an appropriate lens through which to explore the problem of fratricide. The notion of DSA negates the ability to utilise the 2007 model of Wilson et al. in this book, which focuses upon shared cognition.

Wilson et al. (2007) posit that shared cognition is a combination of each team member's knowledge regarding both the task and its environment, and that it is shared throughout the team as a result of team processes. The cyclical team processes are said to enable team members to undertake effective communication, which enables accurate development of shared cognition, which in turn ensures that team members have accurate expectations of one another's roles and can appropriately coordinate with one another (Wilson et al. 2007). Within the concept of shared cognition is the notion that communication and coordination are the processes which allow individual schemata to transform into team cognition, allowing for SA (Wilson et al. 2007). Although the authors agree with the concepts behind this term it is felt that the two underlying aspects of shared cognition, schemata and SA, are too complex to be described under one heading. Salmon, Stanton, Walker and Jenkins, exploring team SA, cite Sarter and Woods (1991), who hypothesise that:

> great care should be taken to differentiate SA from concepts such as mental models. (Salmon, Stanton, Walker and Jenkins 2009: 9)

Indeed, the differing relationships the two constructs hold with other factors also necessitates their exploration as separate concepts.

In addition to the concept of shared cognition, the Wilson et al. (2007) model is taxonomic in nature. The authors do not dismiss the importance of taxonomic methods – indeed, they are argued to be highly useful in the organisation and categorisation of error (Dekker 2003) – but within this research the focus in on the cognitive mechanisms underlying behaviour and the manner in which they interact. The aim of this book is to focus on the behaviour that occurred and the underlying cognition associated with such behaviour, rather than the observable, surface characteristics, the phenotype, of error, or on the attribution of labels (Woods et al. 1994). While not denying the applicability of the model by Wilson and colleagues to the exploration of fratricide, and its importance to fratricide in the identification of communication, coordination and cooperation, the perspective of this research, and the focus on interactions, calls for a different approach.

Gadsen and Colleagues

In line with research conducted by Gadsen et al. (2008) and Gadsen and Outteridge (2006), this book argues for the need to explore fratricide from a systemic perspective, specifically examining the manner in which causal factors interact. The importance of communication within the model is also supported by the research of Gadsen and colleagues. However, within their research Gadsen et al.

(2008) and Gadsen and Outteridge (2006) utilise the Fratricide Causal Analysis Scheme. Although application of the scheme enables important insights into fratricide, Gadsen and colleagues suggest that it is important to look at in-depth examinations of fratricide as well as the breadth afforded by such taxonomies. In line with their suggestion, this research focuses upon an in-depth exploration of the problem rather than on the application of a taxonomy or classification scheme.

Masys

The F3 model is sympathetic to Masys' (2006) theoretical assertions around fratricide causality, focusing upon communication and the military system as a whole. The research presented here has tried to encompass Masys' proposals and tie them together with other fratricide literature in order to develop a model grounded in these ideas. The notion that systems are complex and must be explored as a whole, including the complex way in which system components interact, is key to the F3 model.

Masys proposed a number of important theoretical bases for fratricide research which are reflected in the research presented here. The overarching focus of Masys' research (2006) was on the manner in which systems are designed and procedures are written, hypothesising about the impact of these factors. Although technological and procedural issues are undeniably important to the study of fratricide, the research presented within this book focuses on the team processes involved in incidents of fratricide.

Conclusion

From the review of a number of approaches to fratricide, a series of core suppositions regarding fratricide causality were explored. The exploration of these suppositions within the wider Human Factors literature identified the applicability of a number of theoretical assertions such as Neisser's Perceptual Cycle model and the theory of DSA. In addition to this, and in line with such models, the utility of exploring the problem from a general systems theory perspective, exploring all levels of the military system, and interactions between them, which contribute to the final act of fratricide, has been illustrated. Previous fratricide literature identified five core suppositions important to the study of fratricide. The research in this chapter presents an amalgamation of theories incorporating these five standpoints, developing a new model to explore fratricide causality.

Although the utility and status of grounded theory is evident, Pidgeon and Henwood argue that:

> it makes no sense to claim that research can proceed either from testing theory alone or from a pure, inductive analysis of data. (1997: 255)

In alignment with this assertion the bottom-up research presented in this chapter was supplemented with a top-down analysis. The F3 model has been drawn out of the literature surrounding fratricide, teamwork, SA, expectations and general systems theory. In addition to relating the model back to the literature, Chapter 3, which follows, will attempt to map the model onto an incident of fratricide in order to further reinforce the findings. The combination of top-down and bottom-up approaches, the mixing of theoretical paradigms, has been successfully applied in other safety-critical domains such as nuclear control rooms (Patrick et al. 2006).

Chapter 2

Application of the F3 Model
to a Case Study of Fratricide

Introduction

A review of the current practice surrounding fratricide has allowed for the development of a theoretical model of fratricide causality, the F3 model. In order to build upon, and develop, current theoretical assertions in the literature, this research aims to practically apply the model to incidents of fratricide, populating it with empirical data. This chapter presents initial validation of the model by undertaking an analysis of an incident of fratricide and outlining an attempt to map the incident onto the F3 model.

Methodological Approach

Case Study Research

The research presented within this chapter and in the book as a whole focuses upon single case studies of decision-making. The use of case studies represents an intensive rather than extensive analysis, an emphasis on depth rather than breadth, on exploring a large number of variables, as opposed to a large number of subjects (Denscombe 2007). Case study research has a long-standing history within the Human Factors domain, particularly within naturalistic studies (Smith and Dowell 2000, Salmon, Stanton, Walker, Baber, Jenkins and McMaster 2008, Farrington-Darby et al. 2006, Stanton, Rafferty et al. 2010). Case study research is defined as the exploration of 'a unit of human activity embedded in the real world; understood in context; which exists in the here and now; that merges in with its context so that precise boundaries are difficult to draw' (Gillham 2000: 1).

The in-depth analysis provided by case studies is argued to be the most appropriate approach to explore rare incidents in unique domains (Yin 2003) such as fratricide. The complexity of the problem of fratricide is such that within the research presented in this chapter, a single case study was chosen to be explored. Single case studies allow for 'the development of concepts; the generation of theory; the drawing of specific implications; and the contribution of rich insight' (Walsham 1995, Darke, Shanks and Broadbent 1998: 79). It is due to the complex nature of fratricide that an in-depth evaluation of a smaller sample size was deemed appropriate. In order to explore a larger sample size the analysis would have had a far lower fidelity of information.

Comparison of Actual and Ideal

Previous research into military decision-making has argued against the comparison of performance with formal models (Orasanu 2005, Zsambok 1997), claiming instead that expert performance should be viewed as 'ideal' (Orasanu 2005). In line with this supposition, research from the safety domain suggests that 'formal descriptions of work embodied in policies, regulations, procedures, and automation were incomplete as models of expertise and success' (Woods and Hollnagel 2006: 5).

Therefore, within this research, more effective and less effective performances were compared to draw out core performance differences. In addition to the application of the F3 model to an actual incident of fratricide, an idealised version of events was also modelled. Within the ideal scenario, the incident of fratricide could not have occurred, as it represents all of the possible information, communications and tasks that could have occurred, which may have prevented the incident from occurring. The creation and analysis of this ideal scenario provides a benchmark against which to compare the incident of fratricide. Rather than a search for erroneous action, the comparison of the actual and ideal scenarios allows for the identification of all differences between the scenarios: such differences represent possible causal factors. As Woods et al. stated, 'the same factors govern the expression of both expertise and error' (1994: 197). In order to identify causal factors both success and breakdowns must be explored, the actual scenario clearly represents the breakdowns and the ideal represents the success.

An Incident of Fratricide

Timeline of Events

The incident of fratricide described in the Introduction, in which two British Challenger II tanks engaged one another, occurred in Basra during Operation Telic. Operation Telic began in early 2003 and during the initial stages of the operation 7th Armoured Brigade was involved in the protection of a number of strategic bridges over the Shatt al Basra canal. The focus of this case study is on two of the Battle Groups within 7th Armoured Brigade (the Royal Regiment of Fusiliers (1RRF)) and the 1st Battalion, the Black Watch (1BW), and two specific squadrons within these Battle Groups (C Squadron, Queens Royal Lancers (C Squadron QRL)), which was a part of 1RRF Battle Group, and Egypt Squadron, 2nd Royal Tank

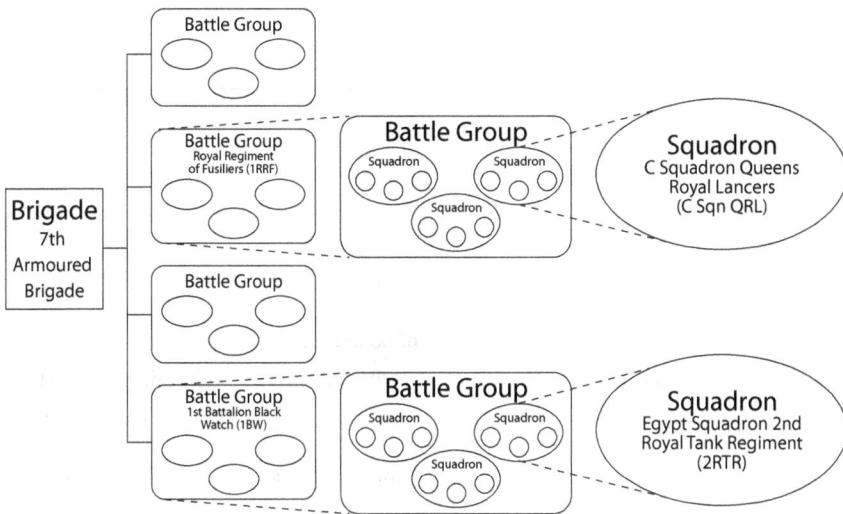

Figure 2.1 Structure of the 7th Armoured Brigade (Ministry of Defence 2004)

Regiment (2RTR) which was a part of 1BW Battle Group) (Ministry of Defence 2004). The structure of the Brigade is illustrated in Figure 2.1.

The 7th Armoured Brigade was deployed in February 2003, beginning operations in southern Iraq in March 2003. The Brigade was tasked by Divisional Head Quarters (HQ) to protect four key bridges over the Shatt al Basra canal. 1RRF (specifically C Squadron QRL) began this protection with a relief in place conducted by 1BW (specifically Egypt Squadron 2RTR) on the 23rd March, in which the task to protect bridge 4 was passed to Egypt Squadron (2RTR). During this handover, confusion arose over Battle Group boundaries and as a result of deficient information about a dam located 1,400 metres north of bridge 4, Egypt Squadron (2RTR) and C Squadron (QRL) became located either side of the canal from one another, unaware of the others' presence. C Squadron (QRL) in position to protect the dam and Egypt Squadron (2RTR) in position to protect bridge 4. Due to the lack of information dissemination regarding friendly force locations (and a myriad other issues), when a Challenger II tank commander from Egypt Squadron (2RTR) noticed movement across the canal, he assumed it to be enemy movement. What the commander identified as enemy personnel entering and exiting a bunker were, in fact, friendly personnel of C Squadron (QRL) entering and exiting another Challenger II as they guarded bridge 4 (Ministry of Defence 2004).

Table 2.1 presents a summary of the key stages of the incident derived from the official government report (Ministry of Defence 2004):

Table 2.1   Key stages of the incident

Event
1BW took over responsibility of bridge 4 from 1RRF
The boundaries of responsibility were changed but information was not adequately disseminated
The dam was not included on Battle Group maps or traces and its presence was not widely known
Battle Group maps were ambiguously marked leading to 1BW and 1RRF developing conflicting beliefs about boundary locations
1BW and 1RRF did not confirm the location of boundaries with one another
The location of two C Squadron (QRL) tanks at the dam was not widely disseminated
Egypt Squadron (2RTR) were unaware of the location of C Squadron (QRL) tanks at the dam
Egypt Squadron (2RTR) was disoriented and confused about the area of main enemy threat
A tank commander from Egypt Squadron (2RTR) observed two hot spots in his thermal imaging equipment in what he believed to be the main area of enemy threat
Egypt Squadron (2RTR) believed there were no friendly forces within 3km
Communication problems meant that Egypt Squadron (2RTR) were unable to contact 1BW HQ
The Egypt Squadron (2RTR) tank commander sent a message to Egypt Squadron HQ asking for information about friendly locations and boundaries
Egypt Squadron HQ were focused on planning an attack and so did not pass the request on to 1BW HQ
The Egypt Squadron (2RTR) tank commander identified the hot spots as enemy personnel
The Egypt Squadron (2RTR) tank commander fired a HESH round, followed by a second round six minutes later

Board of Inquiry Report

An official investigation was conducted by the British Government into the incident of fratricide and the results were presented in a Board of Inquiry report (Ministry of Defence 2004). The Board of Inquiry report concluded that breakdowns in three areas had led to the occurrence of the incident of fratricide: Information dissemination; Coordination; and Combat Identification. The specific issues within these areas, as outlined by the Board of Inquiry, are presented in Table 2.2.

The Board of Inquiry report concluded by stating that the outcome of these breakdowns was a misunderstanding about the enemy status of C Squadron (QRL), a breakdown of SA or what the Board of Inquiry call 'an erroneous appreciation of the situation' (Ministry of Defence 2004: 1). The subsequent report exonerated the commander who fired the shot of any blame; he acted in line with Standard

Table 2.2   Board of Inquiry findings

Area	Breakdown
Information Dissemination	Battle Group boundaries
	Friendly force location
	The presence of the dam north of bridge 4 (key tactical feature)
Coordination	Battle space coordination
	Cooperation among Battle Groups
	Unity of effort within Brigade
Combat Identification	Situation Awareness
	Target identification
	Handover procedures

Operating Procedures (SOPs) based upon the available information that he had. It could be argued that the tank commander himself made no error: he made the correct decision based on the information he had, which in turn was based on his SA. Unfortunately, as the report revealed, this SA was incorrect and thus led him to take action which, with hindsight, was judged to be erroneous.

F3 Model Analysis

Factors

The F3 Model was applied to the example incident of fratricide in order to ensure its ability to explore the causality involved. Analysing the incident of fratricide from the perspective of these Famous Five factors allows a greater understanding of why things went wrong to. Explaining the scenario through errors in Communication, Coordination, Cooperation, Schemata and Situation Awareness provides an understandable causal pathway. Table 2.3, on the following page, illustrates the way in which the five causal factors can be mapped onto the key stages of the incident.

Prior to the incident 1BW conducted a relief in place to take over responsibility for bridge 4 from 1RRF. At the same time as the relief in place occurred, the boundaries of responsibility around bridge 4 were changed to include responsibility for a dam 1,400 metres north of the bridge. Unfortunately, information regarding this boundary change was not widely disseminated, adequate *communication* regarding the boundary change did not occur, resulting in poor SA. Alongside a lack of information transfer regarding the presence of the dam, 1BW had an inaccurate *Situation Awareness* regarding their area of operations and no knowledge of a key tactical feature – the dam.

Table 2.3   Key stages of the incident with assigned key factors

Event	F3 Factor Breakdown	F3 Factor
1BW took over responsibility of bridge 4 from 1RRF		Coord.
The boundaries of responsibility were changed but information was not adequately disseminated	Comm.	
The dam was not included on Battle Group maps or traces and its presence was not widely known	SA	
Battle Group maps were ambiguously marked leading to 1BW and 1RRF developing conflicting beliefs about boundary locations	SA	
1BW and 1RRF did not confirm the location of boundaries with one another	Coord.	
The location of two C Squadron QRL tanks at the dam was not widely disseminated	Comm.	
Egypt Squadron (2RTR) were unaware of the location of C Squadron (QRL) tanks at the dam	Coord.	
Egypt Squadron (2RTR) was disoriented and confused about the area of main enemy threat	Schema	
A tank commander from Egypt Squadron (2RTR) observed two hot spots in his thermal imaging equipment in what he believed to be the main area of enemy threat		
Egypt Squadron (2RTR) believed there were no friendly forces within 3km	Schema	
Communication problems meant that Egypt Squadron (2RTR) were unable to contact 1BW HQ	Comm.	
The Egypt Squadron (2RTR) tank commander sent a message to Egypt Squadron HQ asking for information about friendly locations and boundaries		Comm. SA
Egypt Squadron HQ were focused on planning an attack and so did not pass the request on to 1BW HQ	Coop.	
The Egypt Squadron (2RTR) tank commander identified the hot spots as enemy personnel	SA	
The Egypt Squadron (2RTR) tank commander fired a HESH round, followed by a second six minutes later	Schema	

In addition to the lack of information regarding boundary changes, the maps received by both Battle Groups were not clearly marked and led to conflicting interpretations being made by 1RRF and 1BW, with both holding an inaccurate

Situation Awareness of boundaries. A lack of *coordination* between the two Battle Groups meant that these boundary conflicts were not resolved.

Further problems in *communication* meant that information regarding the location of two C Squadron (QRL) tanks at the dam was not received by Egypt Squadron (2RTR) and the lack of the *coordination* between the Battle Groups meant that Egypt Squadron (2RTR) were unaware of the operations C Squadron QRL were involved in.

The commander of the (2RTR) tank that fired had developed an incorrect *schema* regarding the area of main enemy threat. When the commander noticed hot spots in what he believed to be the main area of threat, this faulty *schema*, along with the *schema* that there were no friendly forces within 3km, meant that he initially assumed the hot spots were indicative of enemy personnel.

The commander was unable to contact his 1BW HQ due to *communication* problems and so sent a request to Egypt Squadron HQ to ask for additional information regarding boundaries and friendly force location. Egypt Squadron was planning an imminent attack manoeuvre and failed to pass on the message to 1BW HQ – a breakdown in *cooperation*.

Consequently, the Challenger II tank commander received no additional information about boundaries or friendly troop positions, despite his request, and was forced to make a decision based upon the information he already had. Each of these breakdowns reinforced the commander's initial assumption that there were no friendly troops within a 3km area and that the hot spots were enemy. This led to incorrect SA that the movement he saw was enemy and thus he fired.

The 'Famous Five' factors could be mapped onto both the key factors of the incident as stated by the government report, and the key factors identified by the analysis undertaken here. Table 2.4 below has been developed to illustrate the ability to map the Famous Five onto the key factors identified in the government report.

Table 2.4   Board of Inquiry report and F3 model factors

Board of Inquiry findings		F3 Model
Area	Breakdown	F3 Factors
Information Dissemination	Battle Group boundaries	Communication/ Coordination
	Friendly force location	Communication/ Coordination
	The presence of the dam north of bridge 4 (key tactical feature)	Communication/ Coordination
Coordination	Battle space coordination	Coordination
	Cooperation among Battle Groups	Cooperation
	Unity of effort within Brigade	Cooperation
Combat Identification	Situation Awareness	Situation Awareness
	Target identification	Schemata
	Handover procedures	Schemata

Development of Ideal Scenario

Using the Board of Inquiry report a timeline of events was built up in which the incident of fratricide could not have occurred. The ideal version of the scenario was created by replacing all inappropriate actions and in the actual scenario, with appropriate replacement actions, as well as inserting any missing actions. The Board of Inquiry report contained sections specifically focused on identifying the actions that should have occurred, with sections titled 'Could this incident have been prevented?' (Ministry of Defence 2004: 35). In these sections the Board of Inquiry report clearly states actions, communications and beliefs that were faulty and identifies the roles, tasks and actions that should have occurred. This allowed for the systematic development of the 'ideal' scenario.

One of the causal factors in the incident was that 1BW (2RTR) were unaware that new Battle Group boundaries had been arranged. Below is an example of the report explicitly stating a faulty action; and highlighting the correct action:

> The 7 Armd Bde trace dated 23 Mar 03 showed 1BW BG's left boundary (and so B Coy's left boundary) as being 500m to the north west of Bridge 4 and ending at the Shatt al Basra Canal. This new boundary would come into effect on 24 Mar 03. As OC B Coy had not seen the latest Bde trace depicting the new boundaries he made an assumption based on his interpretation of previous traces. (Ministry of Defence 2004: 5)

The report explicitly stated that the incident could have been prevented if 'Boundaries had been fully and accurately briefed and the Bde Trace dated 23 Mar 03 been disseminated in a timely fashion' (Ministry of Defence 2004: 35). The report's reference to actions that should not have happened, and its discussion of the ideal actions which should have occurred in their place, allowed for the development of the 'ideal' scenario as presented below in Table 2.5.

Interactions

As the model is concerned primarily with the relationships between the factors and how these relationships break down, an analysis was undertaken to explore the difference in interactions between factors that occurred in the actual and ideal scenarios. The detailed timelines of events created for both the ideal and actual scenario were subject to in-depth coding. Each of the 79 events on the time line was allocated a code based upon the action it described. A detailed coding scheme was developed as presented in Table 2.6. This coding scheme was derived from the literature explored in Chapter 1. A set of instructions was created and strictly adhered to, in order to guide coding.

Table 2.5   Ideal scenario

Event
1BW took over responsibility of bridge 4 from 1RRF
The boundaries of responsibility were changed and information was not adequately disseminated
The dam was included on Battle Group maps or traces and its presence was widely known
Battle Group maps were clearly marked leading to 1BW and 1RRF developing matching beliefs about boundary locations
1BW and 1RRF confirmed the location of boundaries with one another
The location of two C Squadron (QRL) tanks at the dam was widely disseminated
Egypt Squadron (2RTR) were aware of the location of C Squadron (QRL) tanks at the dam
Egypt Squadron (2RTR) appreciated the area of main enemy threat
A tank commander from Egypt Squadron (2RTR) observed two hot spots in his thermal imaging equipment in what he knew was not the main area of enemy threat
Egypt Squadron (2RTR) was aware that there were friendly forces within 3km
Egypt Squadron (2RTR) contacted 1BW HQ to ask for confirmation on boundaries and friendly force locations
1BW HQ sent back confirmation of boundaries and friendly force locations
The Egypt Squadron (2RTR) tank commander identified the hot spots as C Squadron (QRL) tanks

Table 2.6   Coding scheme

Relationship	Rule for coding
Communication – SA	A communication that sends a message about the current environment
Communication – Schema	A communication that sets an expectation about the future
Communication – Cooperation	A communication that causes willingness to undertake behaviour beneficial to the team above normal work responsibilities
Communication – Coordination	A communication that leads to synchronisation and integration of team actions
SA – Communication	An awareness of the situation that leads to a communication
SA – Coordination	An awareness of the situation that leads to synchronisation and integration of team actions
SA – Cooperation	An awareness of the situation that causes a willingness to undertake team behaviour above that required by work responsibilities
SA – Schema	An awareness of the situation that sets an expectation about the future

Table 2.6   Continued

Relationship	Rule for coding
Schema – SA	An expectation that leads to an awareness of the environment
Schema – Communication	An expectation that leads to a communication
Schema – Cooperation	An expectation that leads to a willingness to undertake team behaviour
Schema – Coordination	An expectation that leads to synchronisation and integration of behaviour
Cooperation – Communication	A willingness to undertake coordinated behaviour that leads to a transfer of information
Cooperation – Coordination	A willingness to undertake team behaviour that leads to a transfer of information
Cooperation – SA	A willingness to undertake team behaviour that leads to an awareness of the situation
Cooperation – Schema	A willingness to undertake team behaviour that leads to an expectation
Coordination – SA	Synchronisation and integration of behaviour that leads to an awareness of the situation
Coordination – Schema	Synchronisation and integration of behaviour that leads to an expectation
Coordination – Communication	Synchronisation and integration of behaviour that leads to a communication
Coordination – Cooperation	Synchronisation and integration of behaviour that leads to a willingness to undertake team activity

An illustration of the manner in which the codes were assigned to the mission events is presented in Table 2.7, each representing broken, negative, links in the actual (fratricide) scenario.

The application of the coding scheme to the ideal and actual scenario enabled a matrix of positive links and negative links to be created for both the actual and ideal scenario as illustrated in Table 2.8, Table 2.9 and Table 2.10 on the following pages.

As the ideal scenario is representative of a scenario in which all inappropriate actions are replaced with appropriate actions there are no breakdowns or negative links between the factors. This is not reflective of real life incidents, but it does begin to show which factor interactions are most important by exploring the divergence between positive links in the incident of fratricide and the 'ideal' version.

Using Social Network Analysis, graphical illustrations were produced of the relationships between the factors, positive and negative, in the same manner as the illustrations were derived in Chapter 1. The illustrations are presented in Figure 2.2.

Table 2.7   Example of negative coding in the actual scenario

Event	From	To
Target Identification (ID) process not completed	Schema	SA
Incorrect ID of C Squadron (QRL) tank as enemy in bunker	Schema	SA
Incorrect ID of C Squadron (QRL) tank as enemy vehicle	Schema	SA
Incorrect belief of target location	Communication	SA
Incorrect beliefs about boundaries	Coordination	Schema
No accurate boundary information passed	Communication	Coordination
No accurate information of inter BG boundary	Communication	Coordination
Incorrect beliefs about inter BG boundaries	Coordination	Schema
C Squadron QRL tanks not marked on maps	SA	Coordination
Poor communication between Egypt Squadron 2RTR and 1BW	Communication	Coordination
No dam marked on Battle Group maps	SA	Schema
Change in dam boundary not communicated	Communication	Coordination
1BW did not pass on information about C Squadron (QRL) tanks	Communication	Coordination
Presence of C Squadron (QRL) tanks not widely known	Coordination	SA
Egypt Squadron (2RTR) belief that there were no friendly for 3km	Coordination	Schema
No target reference points coordinated	Communication	Coordination
No grid passed in any reports on target	Communication	Coordination
Message from Egypt Squadron (2RTR) not passed to 1BW by Egypt Squadron	Cooperation	Communication

Table 2.8   Actual scenario positive links

Positive	Communication	Coordination	Cooperation	Schema	SA
Communication		17			
Coordination				1	4
Cooperation		2			
Schema					
SA		2			

Table 2.9   Actual scenario negative links

Negative	Communication	Coordination	Cooperation	Schema	SA
Communication		11		2	5
Coordination				5	1
Cooperation	1				9
Schema	1				13
SA		1		4	

Table 2.10   Ideal scenario positive links

Positive	Communication	Coordination	Cooperation	Schema	SA
Communication		28		2	5
Coordination				6	10
Cooperation	1	2			4
Schema	1				13
SA		3		4	

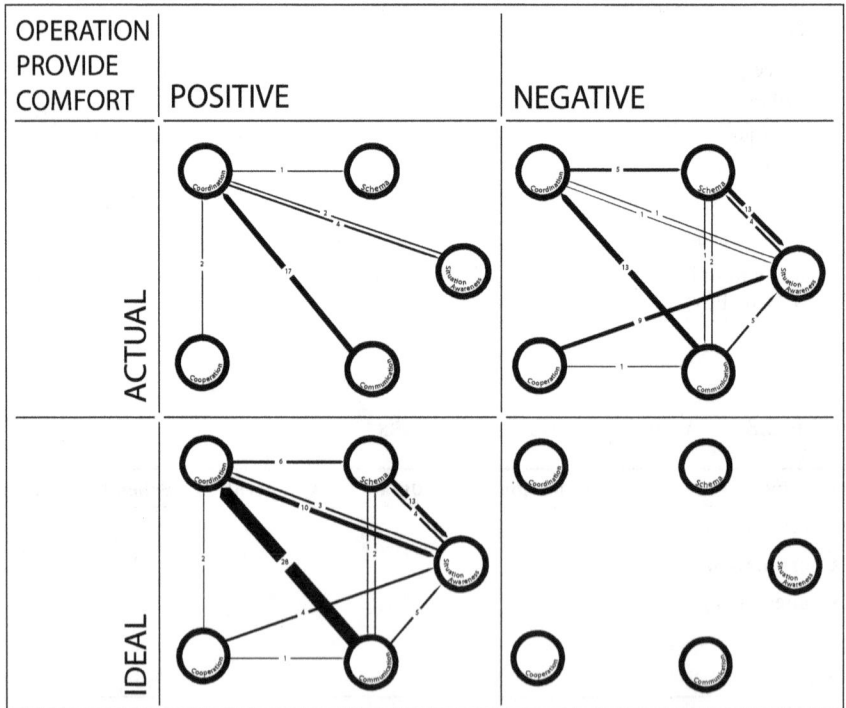

Figure 2.2   Positive and negative links for the actual and ideal scenarios

The illustrations clearly depict which relationships are most important to effective performance, and those relationships which, when broken down, strongly contribute to the occurrence of an incident of fratricide. Within the actual scenario the most important link with respect to breakdowns, the most prominent link, is between schemata and SA. This indicates that the Challenger II tank commander had a high degree of incorrect schemata, for example the notion that there were no friendly forces within a 3km radius, and that these schemata had a large impact on the SA he developed, the pinnacle of this being that the movement he saw through his thermal imaging sights was enemy.

Exploration of the positive links between factors demonstrates that the most prominent link, the most frequently occurring link, is between communication and coordination – although this link is at a much higher frequency within the ideal scenario. From this it could be suggested that although levels of communication between the Battle Groups, and within Battle Groups, did exist within the actual scenario, the levels at which these occurred where not large enough to enable an appropriate level of coordination to occur within the Brigade. That is to say, there were not enough communications to lead to synchronisation and integration of team actions.

Conclusions

Chapter 1 presented the initial development of model of fratricide causality – the F3 model. This model was then applied to an incident of fratricide in order to verify and validate the causal categories. Through the application of the F3 model's causal factors to the authors' own analysis of the incident, as well as the findings from the Board of Inquiry report, it has been illustrated that the key causal factors at play within this incident can be mapped on to the F3 model's causal factors of Communication, Cooperation, Coordination, Schemata and SA. At any time within this accident scenario, adequate levels of communication, cooperation and coordination could have created correct schemata and SA and prevented the shooting of friendly troops. All troops, and specifically the Challenger II tank commander, would have known that the movement detected was friendly personnel.

In addition to this coding-based modelling, the development of social networks enabled an examination of the manner in which the factors interacted with one another in the accident aetiology and in an ideal version of events. The results of this revealed that inappropriate schemata had a dramatic impact on the SA developed, causing a high level of inaccurate SA to be formed. The link between communication and coordination was identified as the most prominent with respect to positive links, suggesting that within the actual scenario not enough communication occurred to ensure an appropriate synchronisation of actions.

The F3 model was able to explain the incident both with respect to the Board of Inquiry findings and exploration of the key stages of the incident within this research. The model encompasses three forms of validity: construct validity (its foundation within the existing literature of fratricide, teamwork and error); face validity (its correlation with existing teamwork, error and fratricide models); empirical validity (initial data from a case study of fratricide fits the model). Such a model is required in order to drive research into the concept of fratricide in an attempt to reduce the occurrence of these incidents.

In conclusion, the previous chapter explored the theoretical literature and developed a prototypical model of fratricide rooted in data and reinforced by previous research into fratricide. This chapter has ensured the ability of the model to explain an example incident of fratricide, and illustrate divergence with 'ideal' performance. Now that the causal factors have been identified and a model created, a method is needed in order to explore these causal factors within incidents of fratricide. Such a method must provide a discussion of each of the causal factors identified within the F3 model.

Chapter 3

FEAST: Fratricide Event Analysis of Systemic Teamwork

Introduction

Chapter 1 identified the need for a systemic exploration of the problem of fratricide. A model was developed from a synthesis of fratricide and wider ergonomics literature, continuing an existing emphasis on the teamwork aspects of fratricide (Wilson et al. 2007, Zobriach et al. 2007). In Chapter 2, this model was applied to an incident of fratricide, with construct validity provided by the model's ability to explore the causal factors identified by the government report into the 2003 Challenger II fratricide incident in Basra. In order to verify and validate the F3 model further, an empirical method is needed to investigate fratricide incidents and the associated causal factors systematically. The investigation of methods which may be appropriate to fratricide investigation begins with an exploration of methods which have previously been utilised in fratricide research.

Previous Fratricide Research

Research into fratricide has employed a number of different methods such as behavioural markers (Jarmasz et al. 2009); Hierarchical Task Analysis (HTA; Jarmasz et al. 2009, Mistry et al. 2009); encounter models (Dean and Handley 2006); classification schemes (Gadsen et al. 2008); and accident analysis methods (Leveson 2002, 2004, Leveson, Allen and Storey 2002), all of which appear to be both appropriate and effective methods to study incidents of fratricide. The aim of this review is to identify the most appropriate approach for a systemic analysis of fratricide in line the theoretical suppositions of this research.

Behavioural Markers

The utility of behavioural markers in the exploration of phenomena is prominent throughout the Human Factors literature (Yule, Flin, Maran et al. 2008). Jarmasz et al. (2009) developed a behavioural rating scheme for fratricide causality based upon the framework of Shared Cognition breakdowns of Wilson et al. (2007). They conducted a HTA of a Close Air Support (CAS) mission in order to identify when these behavioural markers should be measured, and developed a rating scheme against which they could be evaluated. Jarmasz et al. (2009) trialled the method during a simulated Joint Force mission between Canada and the United Kingdom and it was found that there was a correlation between subjective ratings and observers' opinions. The behavioural ratings were able to discriminate between mission behaviours that reflected performance differences, for example three problematic incidents received lower ratings then missions that were performed without incident (Jarmasz et al. 2009).

Unfortunately, the application of the behavioural rating scheme identified a number of problems with the methodology. Jarmasz et al. (2009) found that the method was too complex and that it was unfeasible to rate all behaviours that occurred in a large, multinational, operation using this method. Issues were also raised during the trial regarding the way to interpret the ratings (Jarmasz et al. 2009). Jarmasz and his colleagues found that further research was required into the problem in order to develop a behavioural rating scheme which was more appropriate in nature and number. The literature surrounding behavioural markers states that they should be grounded in a foundation of knowledge to ensure that they are context-specific (Yule, Flin, Paterson-Brown and Maran 2006). It is the authors' opinion that empirical investigation into fratricide is not mature enough to provide such a level of knowledge. It is proposed that further, in-depth, exploration is needed into the problem to develop appropriate behavioural rating schemes.

In addition to the methodological problems the behavioural rating scheme faced, the method does not fit with the theoretical stance of the research presented within this book. The method is unable to explore the majority of the suppositions identified in Chapter 1: it does not explore the interactions between causal factors; it does not explore schemata; and it does not explore SA. Behavioural rating schemes are commonly utilised to evaluate the effectiveness of training techniques (Yule, Flin, Maran et al. 2008), and are a valuable method for evaluating performance. The focus of this research, however, is the exploration of underlying causality – the intricate relationships between causal factors. For this a descriptive diagnostic method is required.

Hierarchical Task Analysis

Mistry et al. (2009) also utilised a HTA to explore Combat Identification. They conducted a HTA on the mission performance of pilots during CAS missions conducted within a simulated environment. They tasked pilot and Forward Air Controllers to undertake CAS missions while under observation in order to develop draft HTAs. The observational notes were supplemented with open-ended interviews and communication transcripts from the mission itself. The HTA method proved to be an effective way to summarise the role of both individuals in the mission, and to identify collaborative tasks. The application of HTA provided a high level of insight into the goals and actions involved in CAS missions. Unfortunately, with respect to the suppositions raised in Chapter 1, the method does not provide an explicit exploration of each; for example it does not provide an investigation of situation awareness, although it can provide an analysis of SA requirements (Salmon, Stanton, Walker and Jenkins 2009).

Encounter Models

The HTA conducted by Mistry et al. (2009) was developed in order to provide empirical data to aid in the development of the INCIDER model, initially mentioned in Chapter 1. The INCIDER model is described as both a conceptual model (representing numerous interrelated factors that may impact on a combat identification decision) and an encounter model (representing combat identification encounters):

> The INCIDER model integrates physical representations of sensors and identification friend or foe (IFF) systems with human cognitive and behavioural characteristics to provide a simplified representation of detection and classification processes. (Mistry et al. 2009: 217)

The INCIDER model represents a promising theoretical and analytical tool. However, at this stage in development the model is focused on the individual decision-maker (work is under way to represent teams within the model) and so is not within the theoretical scope of the research presented within this book.

Classification Scheme

Also mentioned within Chapter 1 is the research presented by Gadsen and colleagues (Gadsen and Outteridge 2006, Gadsen et al. 2008), including their utilisation of the Fratricide Causal Analysis Scheme whereby 12 categories were employed to classify the core causal factors within incidents of fratricide. As has previously been discussed, the utility of the scheme is that it provides a high-level analysis of a wide array of incidents, allowing for classification patterns to emerge. The deficit of the approach is that it does not enable an in-depth analysis of the factors that may be involved, or the manner in which they interact. Indeed, Gadsen and his colleagues themselves suggest that it is important to look at in-depth examinations of fratricide, as well as the breadth of analysis afforded by such taxonomies (Gadsen et al. 2008).

Accident Analysis Methods

Systems Theory Accident Modelling and Process

Leveson and colleagues (Leveson 2002, 2004, Leveson, Allen and Storey 2002) have presented an analysis of fratricide using the Systems Theory Accident Modelling and Process (STAMP) method. STAMP is an accident analysis technique developed to explore accident causation from a systems theory perspective (Leveson 2004). Leveson argues against traditional even-chain models of accident causation and instead proposes that accident causation should account for processes such as adaptation and emergence that are present within complex systems. The method is based upon the hypothesis that accidents occur due to 'external disturbances, component failures, or dysfunctional interactions among system components' which are not sufficiently constrained or controlled by the system (Leveson 2004: 250). Leveson (2004) theorises that a system is made up of multiple levels which interact in unpredictable ways. In order to prevent an accident occurring sufficient control should be enforced in the system to prevent unsafe evolution. The STAMP method provides an organisational hierarchy (control structure), and taxonomy of causation (the classification of flawed control), which is applied to each level of the hierarchy. The application of the taxonomy to the hierarchy identifies where the dysfunctional interactions occurred within the system under analysis.

The method provides an exploration of multiple layers and begins to explore the interaction between them. Leveson (2004) argues that the STAMP control flaws classification scheme provides a number of different levels of analysis which are capable of exploring accident causation at a number of stages of abstraction. The lowest level of the STAMP taxonomy describes that physical chain of events; the next level up explores the conditions surrounding these events; the third level provides an exploration of weaknesses within the system as a whole that allowed the accident sequence to occur (Leveson 2004). These three subsets appear to look at the same erroneous actions at different stages of their emergence, beginning to explore the development and evolution of accidents.

The STAMP methodology also begins to explore the factors of the F3 model, focusing on communication flaws and inaccurate process models (internal cognitive structures) that affect the causality of an accident. The research of A.H. Dekker (2002), S. Dekker (2003), Woods et al. (1994) and Wilson, Salas and Andrews (2009) emphasises the need to explore the accident scenario and understand everything that happened, rather than just re-label error as, for example, communication error. Greitzer and Andrews argue that 'identifying these factors as contributing does not by itself illuminate diagnostic factors underlying these failures' (2009: 174). Providing a description of communication flaws that occurred within the scenario under analysis does not enable an exploration of why these arose. The STAMP methodology begins to explore this causality, for example highlighting inaccurate models that lead to inappropriate communications, illustrating the way in which individuals' expectations may have impacted on the decisions they made within the accident scenario. For the exploration of the F3 model a method is needed that continues this focus but is able to provide explicit exploration of all five of the F3 model factors. It is also suggested that further guidance is required on how to explore complex causality. According to Almeida and Johnson, although the STAMP method highlights prominent decision-makers and decision-making groups, the method:

provide(s) little guidance for any subsequent analysis that might examine the reasons why those groups acted in the manner that they did. (2005: 21)

AcciMap

The utility of the STAMP accident analysis method illuminated other systemic methods as possible candidates to explore fratricide causality. AcciMap (Rasmussen and Svedung 2000) is an accident analysis that employs graphical illustrations to depict accident aetiology. Svedung and Rasmussen (2002), the creators of the method, state that the method is:

> based on the classic cause–consequence chart representing the casual flow of events supplemented by a representation of the planning, management, and regulating bodies contributing to creation of the scenarios. (2002: 403)

The AcciMap creates a visual representation of causal factors and their relationships across six systemic levels. The bottom layer contains the physical factors, the second layer contains the accident process, and the remaining four layers represent a hierarchy of decision-makers who impacted on the accident scenario (Svedung and Rasmussen 2002). Svedung and Rasmussen argue that the aim of the method is to provide a:

> vertical analysis across the levels, not a horizontal generalisation within the individual levels. (2002: 406)

The AcciMap method has a diverse range of applications and has previously been utilised to explore shoot, no shoot decisions within the police domain (Jenkins, Salmon, Stanton and Walker 2010).

AcciMap analysis is able to provide an initial exploration of the relationships between causal factors, both direct and indirect, across the layers of the system. The method moves away from taxonomic labelling focusing on linking causal factors to develop a representation of the accident history. AcciMap is concerned with the local rationality of those involved in accidents, the authors of the method make numerous references to local rationality and hypothesise that the:

> study of decision making cannot be separated from a simultaneous study of the social context and value system in which it takes place and the dynamic work process it is intended to control. (Rasmussen and Svedung 2000: 14)

Yet previous research by Almeida and Johnson (2005) argues that, along with the STAMP method, the AcciMap method provides no guidance on how to explore such local rationality. In addition to this the method does not provide an explicit exploration of the F3 model factors.

Human Factors Classification System

Human Factors Classification System (HFACS) is another accident analysis method that has particular relevance to the military domain, originally developed as an accident analysis tool for the US military (Wiegmann and Shappell 2003). The method is the result of a comprehensive review of accident reports, merged with the theoretical principles of Reason's Swiss Cheese model (Reason 1990) (Wiegmann and Shappell 2003). According to Reason (1990), accidents are the result of breakdowns in four core system layers: unsafe acts; preconditions for unsafe acts; unsafe supervision; and organisational influences. The theory states that the explicit acts involved in an accident scenario, normally labelled as human error, are the result of interacting failures occurring throughout the system, for example problems associated with resource management (Reason 1990). The method utilises the four system layers presented in Reason's model, with a specific taxonomy for each of these layers (Wiegmann and Shappell 2004, Shappell et al. 2007).

HFACS does not provide a depiction of the complex and dynamic relationships between the causal factors involved in an accident scenario. HFACS claims to explore errors across a number of cognitive levels: unsafe acts followed by their preconditions at a number of levels within the system – from physiological conditions such as physical fatigue to psychological conditions such as 'get home it is' to higher levels of the organisation and preconditions such as 'poorly maintained workspaces' (Shappell and Wiegmann 2000: 11). Research by Li, Harris and Yu argues that HFACS explores 'both latent and active failures and their inter-relationships' (2008: 427). Li et al. (2008) were able to find statistical significance in the relationships between errors across the systemic levels e.g. organisational influences to violations. Although the HFACS taxonomy is able to explore different levels, and there may be statistical links between the errors identified by the different levels of the HFACS taxonomy, the identification of causal relationships between these levels is not explored by the HFACS taxonomy.

EAST

Another method which has been employed in accident analysis is EAST (Stanton, Baber and Harris 2008). EAST has been used by numerous researchers in a wide range of domains to analyse complex systems such as Air Traffic Control (Walker, Stanton, Baber et al. 2010), military Command and Control (Stanton, Baber and Harris 2008, Stanton, Jenkins et al. 2009), Naval Command (Stanton, Baber and Harris 2008), Rail Operation (Walker, Gibson et al. 2006) and Police and Fire service emergency response (Houghton, Baber et al. 2006). Recently EAST has been applied specifically to accident analysis. Griffin, Young and Stanton (2010) utilised the method to analyse aviation accident causality, arguing for the benefit of EAST over methods such as STAMP and HFACS, stating that STAMP and HFACS are taxonomic methods, whereas EAST allows for the:

identification of issues arising from decisions made and the information that is or is not available to the correct person at the correct time thus centralising the concept of SA. (Griffin et al. 2010: 199)

The EAST methodology consists of the integration of individual ergonomics methods, including: Hierarchical Task Analysis (HTA; Annett 2005), Coordination Demand Analysis (CDA; Burke 2005), Communications Usage Diagram (CUD; Watts and Monk 2000), Social Network Analysis (SNA: Driskell and Mullen 2005) and Information Networks (IN; e.g. Ogden 1987). Figure 3.1 below illustrates the manner in which the EAST methods are utilised to explore events.

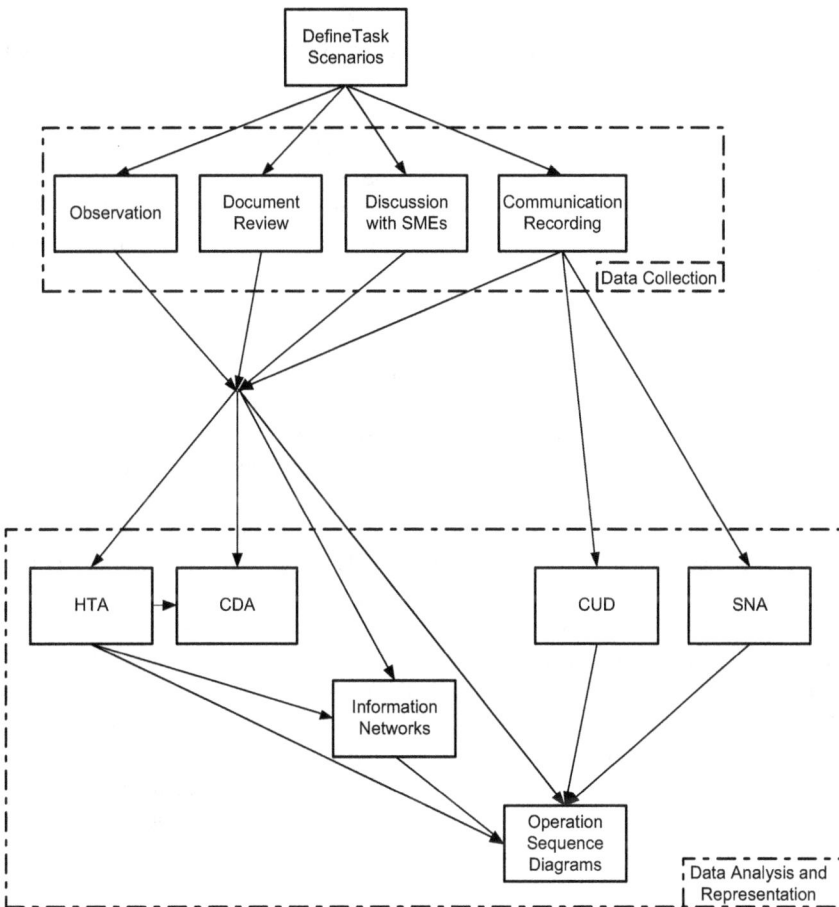

Figure 3.1   EAST methods

Table 3.1   EAST methods applicability (Walker, Gibson et al. 2006)

	HTA	CDA	CUD	SNA	IN
Who					
When					
Where					
What					
How					
Why					

Through the use of multiple methods EAST is able to explore the who, when, where, what, why and how of a scenario (Walker, Gibson et al. 2006), as illustrated in Table 3.1 above.

Walker, Stanton, Baber et al. (2010) explore the utility of EAST in its application to complex sociotechnical systems and emphasise that the combination of methods provided enable multiple perspectives on the exploration of the 'multifaceted' nature of problems such as military Command and Control that a single-method approach could not accommodate. The multiple methods utilised within the EAST methodology also provide a level of validity to the results of the analysis (Stanton Baber and Harris 2008; Walker et al. 2006). Such multiple perspectives are needed in order to explore the emergent properties that arise in complex systems as discussed earlier (Stanton, Baber and Harris 2008).

EAST enables the exploration of quantitative figures to reveal differences and also the contextual description to understand why these differences occur. In this way the method represents a mixture of scientific and anthropological research methods. Walker, Stanton, Baber et al. (2010) argue that experimental methods alone may be unable to capture phenomena emerging within complex systems, suggesting that ethnographic approaches provide a higher level of validity in the exploration of such emergent phenomena. They go on to discuss the limitations associated with ethnographic approaches, including problems associated with generalisation of results, issues with repetition of research, and the fact that such methods are 'couched at a qualitative, often highly discursive level of analysis' (Walker, Stanton, Baber et al. 2010: 186). They put forward EAST as a method that is capable of capturing observable behaviours representative of these complex emergent phenomena; as such, EAST sits between ethnographic approaches, and scientific approaches on an investigatory spectrum, providing the ability to 'reconcile distributed cognition with the methodological traditions of ergonomics' (2010: 187). They conclude that 'the EAST method puts ergonomic analyses in touch with the distributed cognition perspective, rendering the output much more tractable than comparable ethnographic techniques' (Walker, Stanton, Baber et al. 2010: 195), putting forward five main strengths of the EAST method:

1. Focuses on 'objective and manifest phenomena' therefore reducing bias.
2. Is consistent with narrative distributed cognition approaches.
3. Provides results that can be compared across domains.
4. Provides results which are easily summarised and interpreted.
5. Provides a summary which is based upon detailed analysis.

The EAST method is able to align with all of the suppositions presented in Chapter 1. It explores the problem systemically, looking at the interactions between individuals, actions and information. As is pointed out by Griffin, Young and Stanton (2010), EAST provides an explicit discussion of SA and is particularly suited to the exploration of team activities (Stanton, Baber and Harris 2008). In light of these abilities EAST was chosen as the most applicable method to explore fratricide within this book, as the method most closely aligned with the theoretical arguments outlined in Chapter 1.

Application of the EAST Method

The next part of this chapter will apply the EAST method to an incident of fratricide in order to ensure that it is able to practically populate the F3 model with data. In line with the research presented in Chapter 2, exploring the 2003 Challenger II incident in Basra, this section will explore both the actual incident of fratricide and an idealised version of events, in order to identify causal factors.

Incident of Fratricide

Incident Description

The incident of fratricide explored within this chapter occurred during a coalition operation in 1994 called Operation Provide Comfort (OPC) to protect Kurdish refugees in Iraq. Two US F-15 jets were conducting a flight mission to secure a No Fly Zone (NFZ) in the area. During the flight mission the F-15 pilots detected an unidentified radar contact, which they attempted to identify by: contacting the AWACS (Airborne Warning and Control System) team and requesting information on the contact; using IFF technology to interrogate the contact; and conducting visual identification passes of the contact. All three of these attempts led the F-15 pilots to conclude that the contact was a pair of Iraqi helicopters, resulting in the F-15 pilots shooting down the contacts. The contacts were, in fact, two US Black Hawk helicopters (USAF Accident Investigation Board 1994, 1997, Leveson 2002, 2004, Leveson, Allen and Storey 2002, Snook 2000).

The official government report and a number of recent explorations of the incident have provided comprehensive accounts of the causal factors involved (Leveson 2002, 2004, Leveson Allen and Storey 2002, Snook 2000, USAF

Accident Investigation Board 1994, US Government Accountability Office 1997). Leveson (2002, 2004) and Leveson, Allen and Storey (2002) have undertaken a thorough analysis of this incident of fratricide using the STAMP methodology and the reader is referred to these texts for an in-depth exploration of both the fratricide incident and the STAMP methodology. Within this chapter the aim is not specifically to explore the causality involved in the incident; rather, it is to undertake an evaluation of the EAST method applied to the analysis of a fratricide incident from the theoretical stance of this book, identifying its ability to populate the F3 model with empirical data.

A timeline of the incident is presented in Figure 3.2, derived from accounts of the incident (Leveson 2002, 2004, Leveson, Allen and Storey 2002, Snook 2000, USAF Accident Investigation Board 1994, US Government Accountability Office 1997).

Previous analyses of the incident (USAF Accident Investigation Board 1994, US Government Accountability Office 1997, Leveson 2002, 2004, Leveson, Allen and Storey 2002, Snook 2000) have argued that the misidentification occurred because the F-15 pilots had no knowledge regarding the presence of the Black Hawk helicopters. The information about the Black Hawk flights was not included on the daily flight plan. The AWACS team did hold information regarding the presence of the Black Hawks but they did not pass this information on to the F-15 pilots. The Black Hawks were unable to hear communications between F-15 pilots and AWACS, or to communicate with the F-15 pilots, as they were using an incorrect radio frequency for the area. As a result of these factors, when the F-15 pilots noticed radar contacts they assumed they were enemy. An attempt to interrogate the contacts using IFF technology (a system that interrogates other assets to distinguish if they are friendly, unknown or enemy) failed because the Black Hawks were not informed of the correct IFF mode to use within the TAOR (Target Area Of Responsibility), further cementing the notion that the contacts were enemy. The F-15 pilots then conducted a visual identification flight, but due to inadequate visual recognition training, and additional fuel tanks fitted to the Black Hawks, changing their appearance, the Black Hawks were misidentified as Iraqi Hind Helicopters (USAF Accident Investigation Board 1994, US Government Accountability Office 1997, Leveson 2002, 2004, Leveson, Allen and Storey 2002, Snook 2000).

Ideal Scenario

The EAST methodology was applied both to the actual accident scenario and to an ideal version of the accident scenario. In a similar manner to the procedure adopted in Chapter 2, the ideal version of the scenario was created by replacing all inappropriate tasks, communications and information elements in the actual scenario with appropriate replacement actions, as well as inserting any missing actions. Within this ideal scenario, the incident of fratricide could not have occurred. The creation of the ideal scenario was based upon extensive reading around the

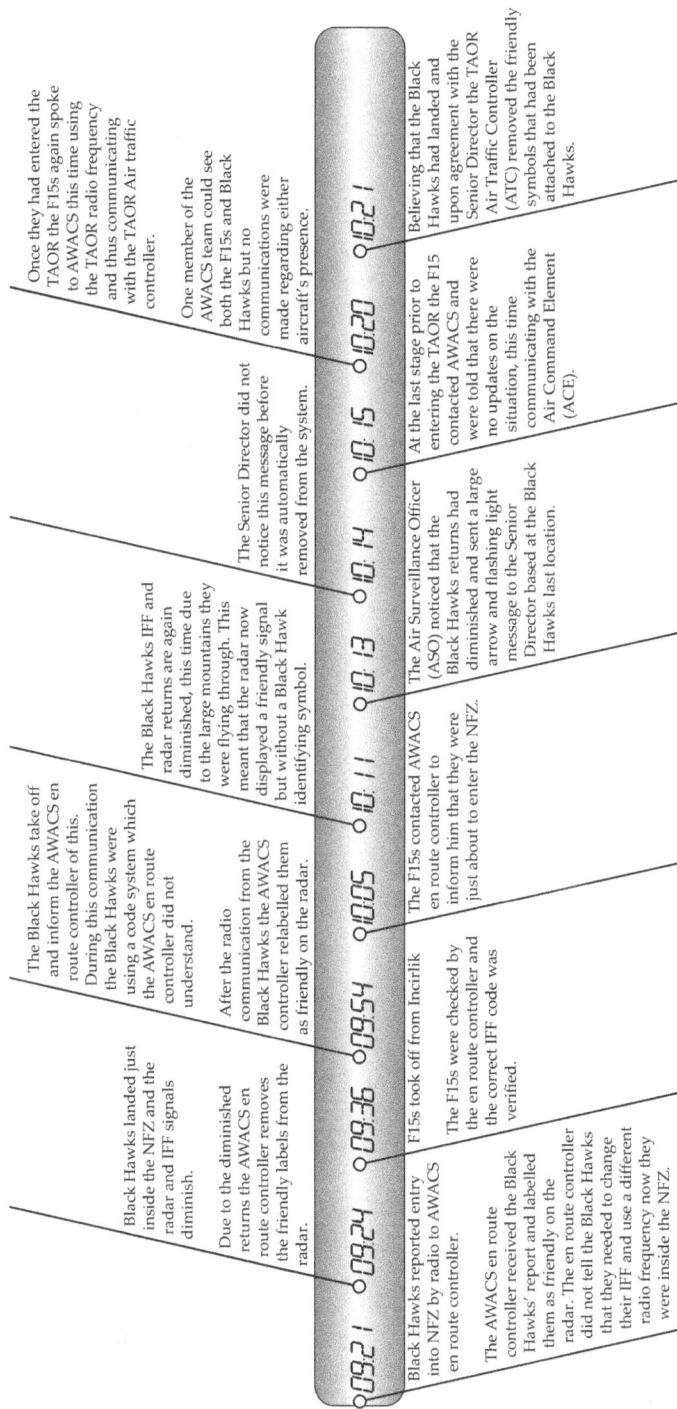

Figure 3.2 Timeline of events

09.21 — Black Hawks reported entry into NFZ by radio to AWACS en route controller.

The AWACS en route controller received the Black Hawks' report and labelled them as friendly on the radar. The en route controller did not tell the Black Hawks that they needed to change their IFF and use a different radio frequency now they were inside the NFZ.

09.24 — Black Hawks landed just inside the NFZ and the radar and IFF signals diminish.

Due to the diminished returns the AWACS en route controller removes the friendly labels from the radar.

09.36 — F15s took off from Incirlik.

The F15s were checked by the en route controller and the correct IFF code was verified.

09.54 — The Black Hawks take off and inform the AWACS en route controller of this. During this communication the Black Hawks were using a code system which the AWACS en route controller did not understand.

After the radio communication from the Black Hawks the AWACS controller relabelled them as friendly on the radar.

10.05 — The F15s contacted AWACS en route controller to inform him that they were just about to enter the NFZ.

10.11 — The Black Hawks IFF and radar returns are again diminished, this time due to the large mountains they were flying through. This meant that the radar now displayed a friendly signal but without a Black Hawk identifying symbol.

The Senior Director did not notice this message before it was automatically removed from the system.

10.13 — The Air Surveillance Officer (ASO) noticed that the Black Hawks returns had diminished and sent a large arrow and flashing light message to the Senior Director based at the Black Hawks last location.

10.14 —

10.15 — At the last stage prior to entering the TAOR the F15 contacted AWACS and were told that there were no updates on the situation, this time communicating with the Air Command Element (ACE).

10.20 — Once they had entered the TAOR the F15s again spoke to AWACS this time using the TAOR radio frequency and thus communicating with the TAOR Air traffic controller.

One member of the AWACS team could see both the F15s and Black Hawks but no communications were made regarding either aircraft's presence.

10.21 — Believing that the Black Hawks had landed and upon agreement with the Senior Director the TAOR Air Traffic Controller (ATC) removed the friendly symbols that had been attached to the Black Hawks.

incident of fratricide and military guides (Leveson 2002, 2004, Leveson, Allen and Storey 2002, Snook 2002, USAF Accident Investigation Board 1994, US Government Accountability Office 1997).

The USAF report clearly states actions, communications and beliefs that were faulty. The report also identifies the roles, tasks and actions that should have occurred. This allowed for the systematic development of the 'ideal' scenario. Below is an example of the report explicitly stating a faulty action; the correct action is also highlighted.

Within the USAF report the erroneous act of the Black Hawks operating with the incorrect IFF frequency for within the TAOR is depicted:

> F-15 pilots attempted to electronically identify the radar contacts by interrogating the ATO-designated IFF Mode I and Mode IV aircraft codes. The helicopter crew members were apparently not aware of the correct Mode I code specified for use within the TAOR and had the Mode I code specified for use outside the TAOR in their IFF transponders. The result was that the F-15s did not receive a Mode I response. (USAF Accident Investigation Board 1994)

The USAF report references changes to the system that have now been introduced to prevent the incident happening again; these are actions that should have occurred but did not. With respect to the above example the 'ideal' is referred to three times in the USAF report:

> All aircraft (including helicopters): Require contact with Airborne Warning and Control System (AWACS) on TAOR Ultra High Frequency (UHF) Have Quick or UHF clear radio frequencies and confirmation of Identification Friend or Foe (IFF) Modes I, II, and IV. If either negative radio contact with AWACS or inoperative Mode IV do not proceed into TAOR.

> Immediately after takeoff contact AWACS and reconfirm IFF Modes I, II and IV are operating.

> All Aircraft (including helicopters): Must be under positive control (i.e., radio contact and positive IFF/SIF) of AWACS to operate inside the TAOR. Require positive IFF/Special Identification Feature (SIF) and radio checks be accomplished while enough fuel remains to return to Diyarbakir AB. (USAF Accident Investigation Board 1994)

Within the USAF report the IFF mode check is clearly stated as the AWACS weapon directors' role:

> One WD (weapons director) acts as an en route controller, responsible for controlling the flow of aircraft to and from the TAOR. This person also conducts IFF and radio checks on all OPC aircraft. (USAF Accident Investigation Board 1994)

The USAF report explicitly refers to actions that should not have happened, and the report's discussion of the ideal actions which should have occurred in their place allows for the development of the 'ideal' scenario.

Criteria

The key aim for this chapter is to identify a method which could be used to explore an incident of fratricide adequately, in line with the theoretical stance and model proposed in Chapter 1. The method must be able to explore the problem of fratricide from a theoretically valid standpoint, in relation both to this book and to the wider literature surrounding safety and accident analysis. Identification of an appropriate methodology entails an exploration of the method's theoretical underpinnings. Dekker highlights the importance of the theoretical underpinnings of a method, stating that:

> measuring without an explicit underlying model that directs observations and allows classification is folk science. (Dekker 2003: 97)

In order to answer this question a set of guidelines was developed to provide a measurement of EAST's performance. The criteria applied in this research have been developed based upon the insights from previous fratricide research and from the literature surrounding accident investigation methods, focusing upon accident analysis research by Qureshi (2007), Sklet (2002), Woods et al. (1994), Hollnagel (2005), Reason (1990) and Dekker (2003). Examination of this literature identified a number of key guidelines.

Qureshi emphasised the importance of two key aims of accident investigation:

1. 'Capture the complexities and dynamics of complex systems' (Qureshi 2007: 10).
2. Capture 'emergent phenomena, which arise due to the complex and nonlinear interactions among system components' (Qureshi 2007: 10).

Sklet (2002, 2004) highlighted a number of components that an accident investigation should include:

1. An accurate and rational description of the events that occurred within the accident scenario.
2. A methodical and systematic framework and arrangement of procedures to guide the analyst.
3. A comprehensive analysis identifying all contributing causal factors.
4. An exploration of all systemic levels involved.

Woods et al. (1994) present a prominent text on the investigation of error from which a number of core hypothesis can be drawn:

1. Erroneous actions represent the beginning of accident investigation.
2. The label of human error is only attributable after an accident has occurred.
3. Accident causality is the evolutionary result of the combination of multiple causal factors.
4. Causality involves multiple, interacting people from throughout the system.
5. Exploration of local rationality is critical.
6. The dynamics and evolution of accidents are important.

Dekker's (2002) ideas on accident investigation methodology are prominent in the literature. The 'new view' of error supported by Dekker sees error as an illustration of trouble inherent within the system. Human error is seen as 'systematically connected to features of people's tools, tasks and operating environments', with safety progress only coming from 'understanding and influencing these connections' (Dekker 2002: 372). Following from his systemic depiction of accident scenarios, Dekker puts forward a series of guidelines for accident investigation from a systemic approach. These include:

1. The exploration of local rationality.
2. The clear illustration of the events leading to the accident.

Hollnagel (2005) emphasised the multi-causal and emergent nature of accidents, hypothesising that 'some properties of the system cannot be attributed to individual components but rather emerge from the whole system' (Hollnagel 2005: 1–6).

The literature explored within Chapter 1 highlighted a number of concepts that were important to the analysis of fratricide:

1. Expectations impact Situation Awareness.
2. Situation Awareness is an important factor of fratricide causality.
3. Teamwork is an important factor of fratricide causality.
4. Fratricide is complex, multi-causal and is the result of problems at multiple levels of the military system.
5. Further research is needed into the interactions between causal factors.

From the work of these researchers, as well as the wider literature and the literature from the fratricide domain, a set of criteria for accident investigation methods was developed. Table 3.2 represents the six criteria that an accident analysis method should meet in order to explore a complex problem such as fratricide, from the theoretical stance of this book.

Table 3.2   Summary of criteria

Criteria	Source
Explore each of the Famous Five factors	Chapter 2
Explore multiple erroneous actions	Kogler (2003), Dean and Handley (2006), Gadsen and Outteridge (2006)
	US Congress (1993), Masys (2006), Wilson et al. (2007), Zobarich et al. (2007)
	Hollnagel (2002), Benner (1985), Kirwan (1992b, 1998a)
	Gadsen et al. (2008), Jamieson and Wang (2007), Wagenaar et al. (1997)
Explore causal relationships	Jamieson and Wang (2007), Dean and Handley (2006)
	Gadsen et al. (2008), US Congress (1993), Masys (2006), Wilson et al. (2007)
	Kogler (2003), Hollnagel (2002), Von Bertalanffy (1950), Qureshi (2007)
	Zobarich et al. (2007), Woods et al. (1994), Gadsen and Outteridge (2006)
Capture all decision makers	Jamieson and Wang (2007), Dean and Handley (2006), Gadsen and Outteridge (2006)
	Gadsen et al. (2008), US Congress (1993), Masys (2006), Wilson et al. (2007)
	Kogler (2003), Sklet (2002), Woods et al. (1994), Zobarich et al. (2007)
Provide clear guidance	Benner (1985), Sklet (2002), Kirwan (1998a), Dekker (2002, 2003)
	Stanton et al. (2005), Stanton and Young (1999)
Capture local rationality	Dekker (2003), Woods et al. (1994), Reason (1990)

Criterion One: Explore Each of the Famous Five of Fratricide Factors

The Famous Five of Fratricide (F3) developed in Chapter 1 incorporates five core causal factors and the interactions between them in order to explore incidents of fratricide. To enable validation and verification of the model, an applicable investigatory method is required to explore fratricide scenarios. The F3 model's core factors are Communication, Coordination, Cooperation, Schemata and Situation Awareness. A method is required which is capable of exploring each of these five factors.

Criterion Two: Provide Comprehensive Exploration of all Erroneous Actions

Previous research into fratricide, discussed in Chapter 1, has highlighted that the problem is one of multi-causality and it is important to explore all of the possible causal factors involved. Sklet (2002) reinforces the importance attached to exploring all causal factors in accident investigations, stating that a comprehensive analysis identifying all contributing causal factors is required. The example incident of the Black Hawk shoot-down in Operation Provide Comfort has been subject to multiple analyses, by numerous researchers, using a variety of methodologies, for example Leveson (2002, 2004), Leveson, Allen and Storey (2002), Snook (2000) and Ladkin and Stuphorn (2003). Throughout these analyses, the seven causal factors below have been present, and it can therefore be assumed that they are fundamental causal factors in the incident. EAST will be rated on its ability to identify each of these causal factors. The seven errors were:

1. The Combined Task Force failed to provide clear guidance leading to an unclear understanding of roles throughout the coalition force, specifically with regard to supporting helicopter missions.
2. The Combined Task Force did not view helicopters as part of the air operations, so that they were not integrated adequately or monitored sufficiently when flying in the TAOR.
3. There was an insufficient level of training given regarding the Rules of Engagement within the TAOR, resulting in a simplified understanding of the ROE.
4. The AWACS team held information regarding the presence of the Black Hawks but did not pass this on to the F-15 pilots.
5. The Black Hawks were unaware of the correct IFF Mode or radio frequency to use inside the TAOR.
6. Due to poor visual identification training, and additional fuel tanks on the Black Hawks, the F-15 pilots misidentified them as enemy.
7. The helicopters were not fitted with contemporary radios and so were unable to talk to the F-15 pilots.

Criterion Three: Provide Exploration of Causal Relationships

The research presented in Chapter 1 emphasised the need to explore fratricide causality in terms of the relationships between causal factors highlighting the emergent and non-linear relationships that may impact causality. Qureshi highlights this point in relation to accident investigation methods, stating the importance of exploring the:

> dynamics and nonlinear interactions between system components in complex socio-technical systems. (Qureshi 2007: 2)

Woods et al. (1994) also emphasise the importance of dynamics and their emergence within accident scenarios, positing that the label of 'human error' should only be the first step, the beginning, of an investigation into accident causality. Woods et al. state that:

> it is the investigation of factors that influence cognition and behaviour of groups of people, not the attribution of error in itself. (Woods et al. 1994: 21)

In line with the core principles of general systems theory, it is the influence of causal factors upon one another that leads to emergent properties (Von Bertalanffy 1950) such as the emergence of an accident (Qureshi 2007). An accident investigation methodology must explore the relationships that exist between causal factors within the accident scenario in order to identify the manner in which the accident emerged.

Criterion Four: Capture all Decision-Makers Involved

Research into fratricide has highlighted the multi-level nature of the problem. The focus should not be on the end decision to shoot but rather on how the military system as a whole evolved to allow the incident to occur (Gadsen et al. 2008, Gadsen and Outteridge 2006). Within the accident investigation domain Sklet (2002) and Woods et al. (1994) emphasise the need to explore individuals throughout the entire system.

Criterion Five: Provide Clear Guidance

The examination of incidents of fratricide is complex and guidance is required in order to ensure that the analysis is comprehensive, identifying all associated causality. Researchers such as Kirwan (1998a), Stanton and Young (1999), Dekker (2002, 2003), Sklet (2002) and Stanton et al. (2005) have highlighted the need for accident investigations to have a rigid structure in order to ensure reliable and valid findings and to 'leave a trace that others can follow' (Dekker 2002: 379).

Criterion Six: Capture Local Rationality

Woods et al. explore the concept of local rationality stating that:

> what people do makes sense given their goals, their knowledge and their focus of attention at the time. Human problem solvers possess finite capabilities. There are bounds to the data that they pick up or search out, limits to the knowledge that they possess, bounds to the knowledge that they activate in a particular context. (Woods et al. 1994: 16)

Dekker builds upon the notion of local rationality and its importance in accident investigation, advocating the need to 'identify people's goals, focus of attention and knowledge active at the time' (2002: 381) in order to understand the individual's perception of the world at the time the accident was occurring and how this influenced their interpretation of the accident scenario.

Application of EAST

The next part of this chapter contains a presentation of the application of the EAST method (Stanton, Baber and Harris 2008) to the incident of fratricide which occurred during Operation Provide Comfort. The analysis is structured around the criteria outlined above and the method's ability to fulfil each of these.

Criterion One: Explore Each of the Famous Five of Fratricide Factors

Through its use of multiple methods EAST provides the ability explore a problem from multiple perspectives. These multiple perspectives align with the F3 model's five factors as illustrated below in Table 3.3.

Table 3.3   EAST method and F3 factors

F3 Model Factor	EAST method
Communication	Social Network Analysis
	Communication Usage Diagram
Coordination	Coordination Demands Analysis
	Hierarchical Task Analysis
Cooperation	Social Network Analysis
	Hierarchical Task Analysis
Schema	Information Networks
Situation Awareness	Information Networks

Coordination and Cooperation

Hierarchical Task Analysis (HTA)

The HTA describes the task or system under analysis in terms of the overall goals, sub-goals and task steps undertaken (Annett 2005, Stanton 2006). HTA is a generic method that can be used to provide an analysis of the goal and task structure used within any domain (Stanton 2006, Walker, Gibson et al. 2006) and insights into

SA requirements (Salmon, Stanton et al. 2009). From the HTA the task network is created which represents the high-level goals and their interactions with one another derived from the detailed task analysis. The task networks provide the ability to compare the manner in which teams broke down high-level goals and the task steps utilised to accomplish these goals.

The full HTA for the two mission scenarios are too large to be presented within this book. Instead summary illustrations of the task networks are presented in Figure 3.3 and Figure 3.4, illustrating the high-level goals and the way in which these are connected.

The task networks immediately illuminate differences in the strategy of mission completion between the two scenarios. Within the actual scenario a linear chain of command was identified with the tasks highly separated and two very separate agencies (CFAC and MCC) performing the different tasks. Within the ideal scenario, however, the structure would be far more networked, with tasks being undertaken collaboratively. This would represent a higher level of collaboration within the ideal system than within the actual system.

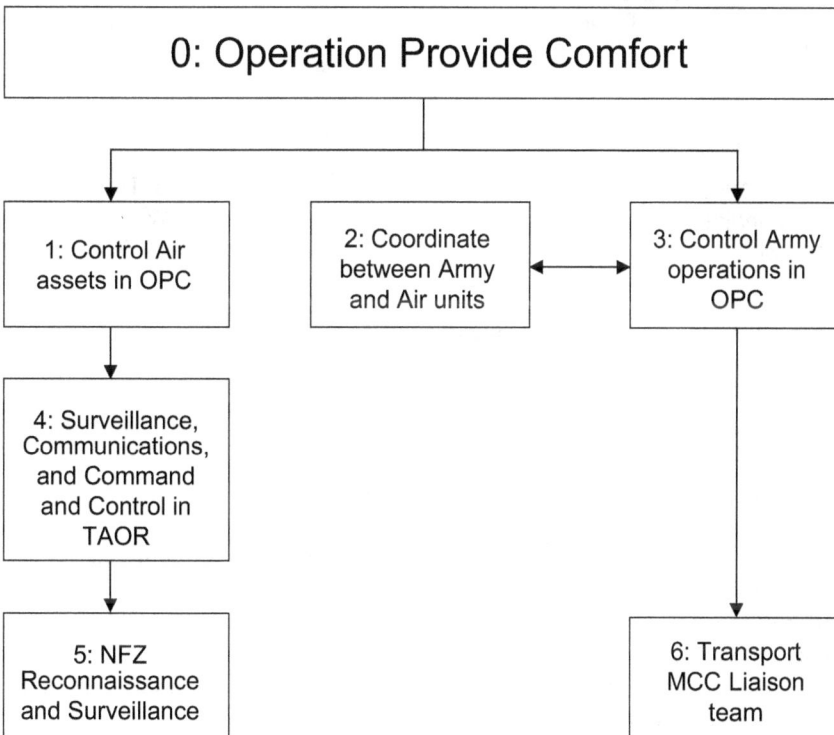

Figure 3.3   Task network for actual scenario

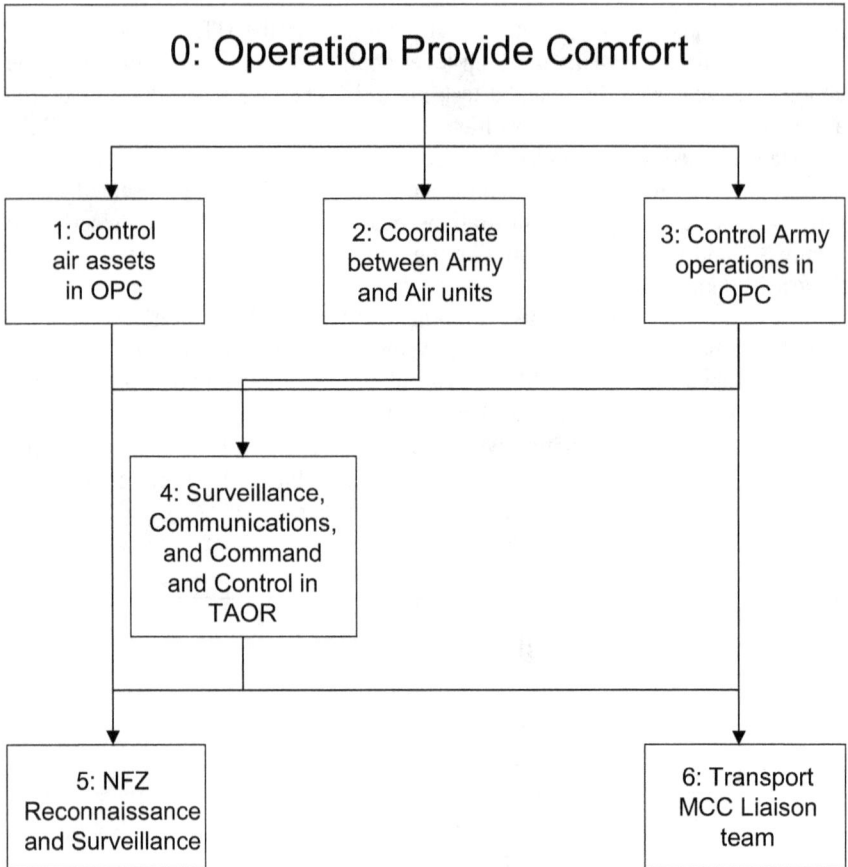

Figure 3.4   Task network for ideal scenario

Coordination

Coordination Demands Analysis

The Coordination Demands Analysis (CDA) provides a discussion of all coordinated activity that occurred during the scenario, as well as rating the effectiveness of this coordination. Tasks are extracted from the HTA and rated as either individual work tasks or teamwork tasks (Burke 2005). Like the HTA, the full CDA is too extensive to be included within this book and so a summary of the results is presented in Table 3.4.

This analysis revealed that within the actual scenario 50 tasks involved individuals and 77 tasks involved teams, compared to 54 individual tasks and 98 team tasks in the ideal scenario. When these results are converted into a percentage

Table 3.4   CDA results

	Individual n (%)	Team n (%)
Actual	50 (39%)	77 (61%)
Ideal	54 (35%)	98 (65%)

of the total tasks undertaken by each team, the results reveal comparable levels of teamwork in both teams.

The next stage of the CDA involves the application of a taxonomy containing communication, Situation Awareness, decision-making, mission analysis, leadership, adaptability and assertiveness (full details of the taxonomy can be found in Burke 2005). The taxonomy is applied to each task step judged to be a teamwork task in the HTA, and each of these tasks is rated low, medium or high (1–3). From the application of this taxonomy a total coordination score can be derived from the mean of the component scores (Stanton, Salmon et al. 2005). The mean scores for the actual and ideal scenarios are presented below in Figure 3.5.

Within the ideal scenario there were higher levels of coordination within teamwork tasks, a total coordination value of 2.7, compared to 1.9 in the actual scenario. According to Stanton, Baber and Harris (2008) a rating of 2.25 or over is representative of a high level of coordination activity, illustrating a high level within the ideal scenario but not within the actual scenario. The results of the CDA analysis identify a similar percentage of teamwork tasks in both scenarios but a greater quality of coordination within the ideal scenario.

As is illustrated within this section, the CDA is able to provide an exploration of coordination. Through the methodology quantitative measurements of coordination can be derived allowing for statistical comparisons between teams or even between scenarios.

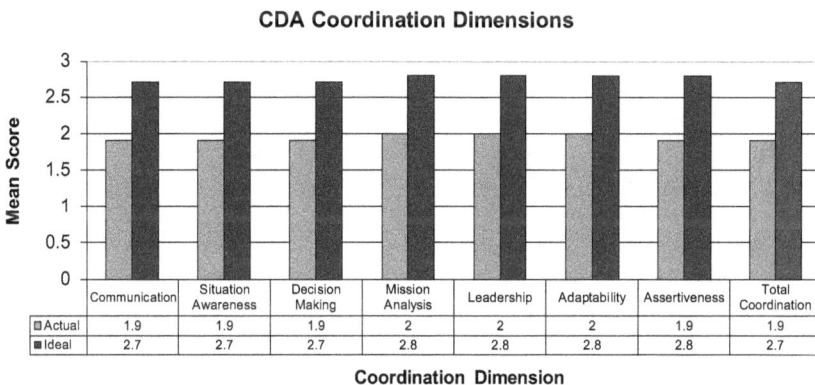

CDA Coordination Dimensions

	Communication	Situation Awareness	Decision Making	Mission Analysis	Leadership	Adaptability	Assertiveness	Total Coordination
Actual	1.9	1.9	1.9	2	2	2	1.9	1.9
Ideal	2.7	2.7	2.7	2.8	2.8	2.8	2.8	2.7

Coordination Dimension

Figure 3.5   CDA dimensions

Communication and Cooperation

Social Network Analysis

Social Network Analysis (SNA) is 'a method for analysing relationships between social entities' based upon an exploration of 'indices of relatedness among entities to represent social structures' (Driskell and Mullen 2005: 58–1). The social network illustrates the communication relationships between individuals within the system based on all communications that occurred and the individuals they occurred between. SNA 'focuses on the relationships among actors embedded in their social context' (Driskell and Mullen 2005: 58–1), ensuring that the method is aligned with the principles of general systems theory (Walker, Stanton, Baber et al. 2010).

The first stage of analysis involved deriving a 'from – to' communication matrix for each agent in both scenarios. The two communication matrices were then inputted into a SNA software tool called AGNA (Applied Graph and Network Analysis; Benta 2003) in order to develop visual representations of the communication links utilised during mission performance. AGNA was also used to derive a number of graph theory metrics to statistically explore these communication relationships as prescribed in previous research (Driskell and Mullen 2005, Walker, Stanton and Salmon 2011, Walker, Stanton, Baber et al. 2010) allowing for direct and simple comparison with other systems' communication structures (Walker, Stanton, Baber et al. 2010).

From the social network matrices the social network communication relationships between individuals in the scenario were developed into graphical representations. Figure's 3.6 and 3.7 present the social networks for the actual and ideal scenarios. The thickness of the lines between actors represents the frequency of communication between them; a thicker line is indicative of a high frequency of communication. The lines are also annotated with the communication frequency data.

Visual examination of the social networks reveals differences between the social organisation in the actual and ideal scenarios. In the ideal scenario three key agents can be identified – CTF, CFAC and F-15 lead – as holding links to the majority of other agents involved in the scenario. The actual accident scenario appears to be more convoluted, with many agents having links to many other agents and a lower level of key, or central, agents with fewer, stronger links. From this it can be surmised that the ideal network is more centrally organised around a number of key agents – a few strong communication links, whereas the actual scenario is more distributed with a greater number of weaker communication links.

Graph theory metrics have been used to explore node and edge relationships across many different domains (Driving: Walker, Stanton and Salmon 2011; Command and Control: Salmon, Stanton, Walker and Jenkins 2009, Stanton, Baber and Harris 2008, Stanton, Jenkins et al. 2009). Within this research five graph theory metrics are discussed: diameter; density; centrality; sociometric status; and cohesion.

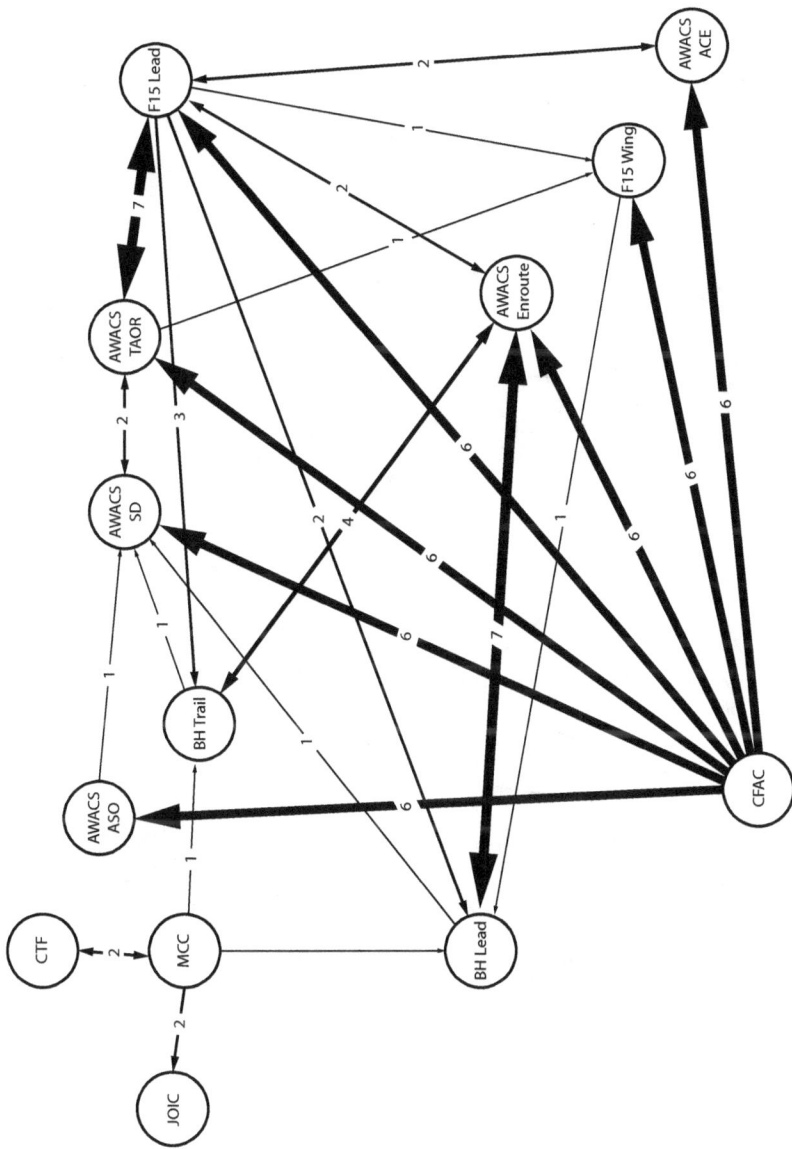

Figure 3.6 Actual scenario social network

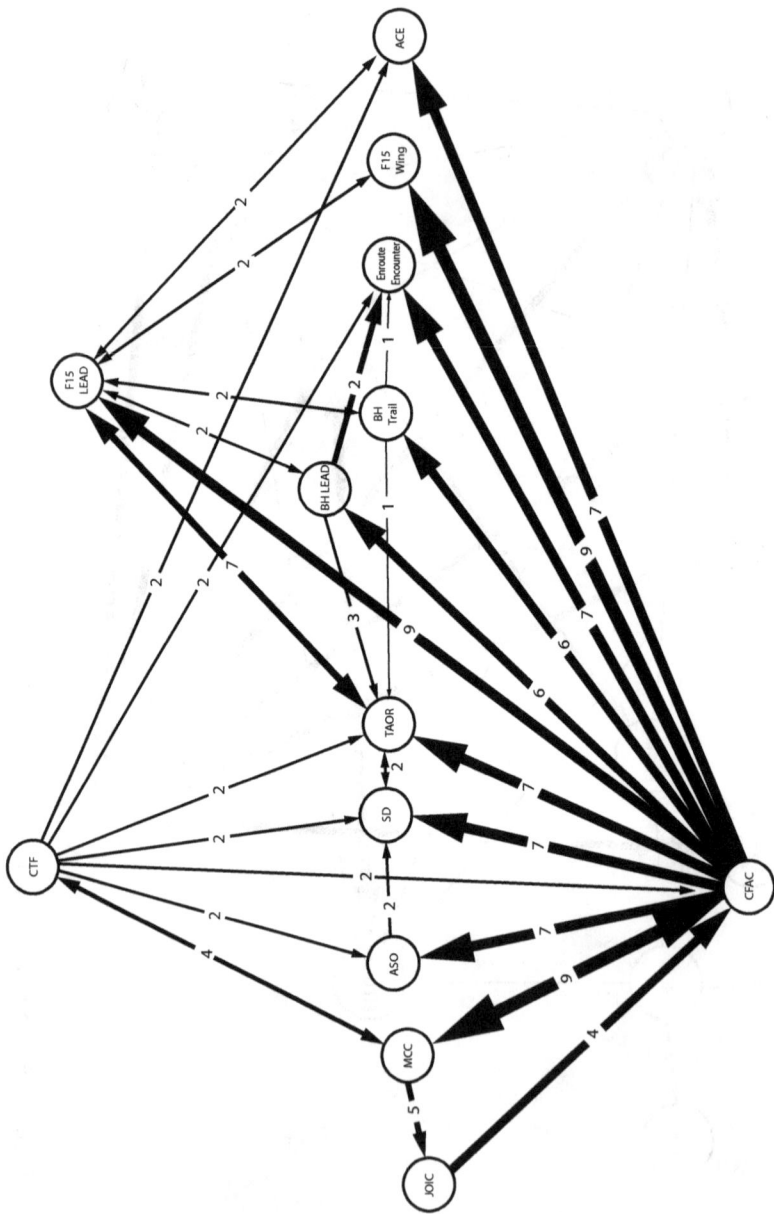

Figure 3.7   Ideal scenario social network

$$Diameter = \max_{uy} d(n_i, n_j)$$

Diameter is a metric to represent 'the largest number of [agents] which must be traversed in order to travel from one [agent] to another' (Weisstein 2008, Harray 1994). Generally speaking, the bigger the diameter, the more agents there are on lines of communication. Hierarchical organisations are proposed to have larger levels of diameter compared to fully connected networks due to the additional layers between sender and receiver.

$$Density = \frac{2e}{n(n-1)}$$

Density represents the level of direct connections between individuals in the team as a whole; a higher level represents a more connected team (Walker, Stanton and Salmon 2011). It is proposed that a hierarchical structure will be characterised by a lower level of density than a fully connected network, suggesting a greater ease of information dissemination in the all-connected network due to the increased level of connections between individuals.

$$Centrality = \frac{\sum_{i=1 \, j=1}^{g} \delta ij}{\sum_{j=1}^{g} (\delta ij + \delta ji)}$$

Centrality represents the prominence of each individual within the team as a whole, based upon how closely they are located to other individuals, how tightly clustered they are (Walker, Stanton and Salmon 2011) and the 'distance' between them (Houghton, Baber et al. 2006). This is a similar measure to diameter but looking at individuals, whereas diameter looks at the team as a whole (Walker, Stanton and Salmon 2011).

$$Sociometric \; status = \frac{1}{g-1} \sum_{j=1}^{g} (xji + xij)$$

Sociometric status represents the level of contribution each individual makes to the communication occurring within the network, or 'how busy' an individual is in relation to other individuals (Houghton, Baber et al. 2006: 1206).

Cohesion represents 'the number of mutual connections in the network divided by the maximum possible number of such connections' (Benta 2003).

These metrics were chosen for examination as they enable an illustration of the connections and clustering of individuals at both a team and individual levels. The results of the analysis reveal a number of differences between the communication networks of the less effective and more effective teams.

Sociometric Status

Examination of the sociometric status data for each individual involved in the mission (see Figure 3.8 below) reveals differences between the actual and ideal scenario. Only one individual within the ideal scenario has a lower status (representing a lower level of contribution to communication) than in the actual scenario – AWACS Enroute controller. This difference can be explained by the role AWACS Enroute controller played in the mission scenario. Within the actual scenario AWACS Enroute controller maintained control of the Black Hawk helicopters throughout their flight when he should have passed control over to the TAOR AWACS controller, thus his increased contribution to the communication within the mission was a result of incorrect actions which contributed to the incident of fratricide occurring (if control of the Black Hawks had been passed to the TAOR controller he would have been aware of both the F-15 jets and the Black Hawks and the incident could have been prevented). Excluding the AWACS Enroute controller, all other individuals within the ideal scenario have a higher level of sociometric status, indicating a higher level of communication contribution compared to the actual scenario.

A dramatic difference between the actual and ideal scenario is illustrated by the spike in sociometric status for CFAC within the ideal scenario. This difference represents a lower than optimum level of communications involving the CFAC within the actual scenario, suggesting that the CFAC should have played a more dominant role in the mission. A greater level of communication from CFAC, who were aware of the Black Hawk flight, with other individuals in the military system could have prevented the incident of fratricide from occurring.

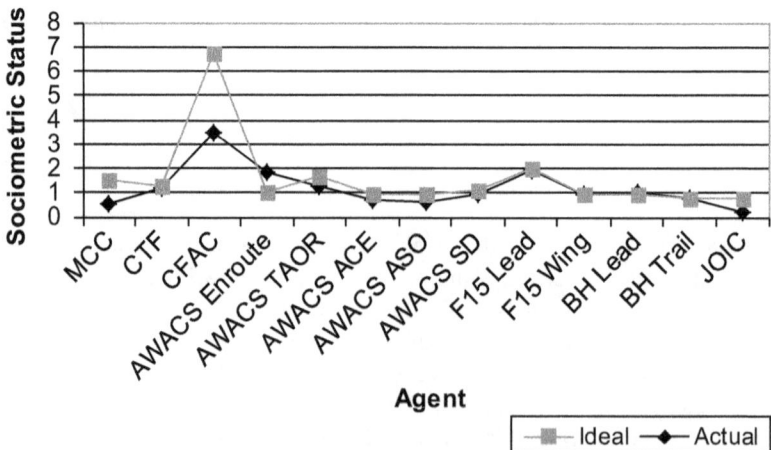

Figure 3.8   Sociometric status

Centrality

The centrality metrics mirror the sociometric status metrics, emphasising the greater level of contribution from higher system levels within the ideal scenario compared to the actual scenario, with both CTF and CFAC holding higher centrality values in the ideal scenario, as shown in Figure 3.9.

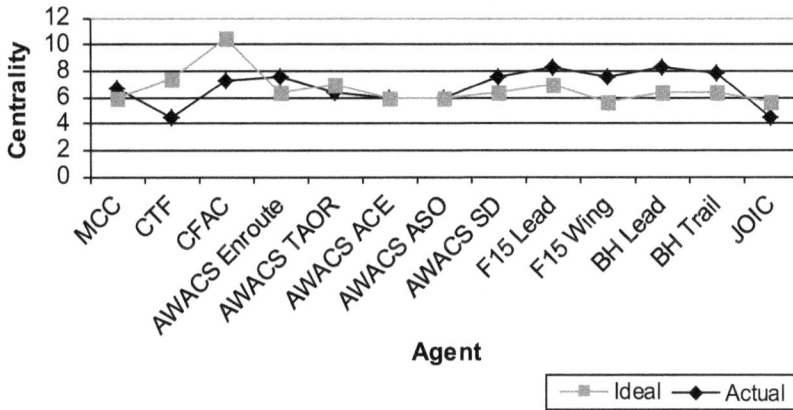

Figure 3.9   Centrality values for actual and ideal scenarios

The results of the analysis of cohesion, diameter and density are presented below in Table 3.5.

Table 3.5   SNA results

	Actual	Ideal
Cohesion	0.1	0.07
Diameter	5	5
Density	0.21	0.23

There is little difference in the levels of interconnectivity within the networks, the density. Individuals held similar levels of communication connections in both the ideal and actual scenarios. The diameter metric also illustrates a similar level of clustering within both scenarios. It appears that the communication structure was similar between both scenarios but the utilisation of structure differed significantly, as is illustrated by the sociometric status analysis.

A higher level of cohesive bonds was found in the actual scenario when compared with the ideal scenario. Cohesion between team members is traditionally accepted to be beneficial. However, in relation to social networks, research has illustrated a negativity attached to high levels of cohesion. For example, it may be that the cohesive bonds within MCC and CFAC were too strong and led to these systemic levels isolating themselves from one another. This could be a causal factor in the poor communication between these two systemic levels. This finding is in line with research from the communication domain; for example Katz et al. (2005) cite work by Markovsky, Heimer and O'Brien (1994), in which higher levels of cohesion found in 'cliques' were said to have a negative impact on the team as a whole.

Situation Awareness and Schemata

Information Networks

The information networks within this chapter were developed from content analysis of the documentation surrounding the incident of fratricide. Content analysis is used to draw out the information elements held by individuals within the system, as well as identifying the causal links between these information elements (Walker, Stanton, Baber et al. 2010). These information elements are 'analogous to propositions, i.e. entity or phenomena about which an individual requires information in order to act effectively' (Walker, Stanton, Baber et al. 2010, Walker, Gibson et al. 2006).

Salmon, Stanton, Walker and Jenkins (2009) present a set of steps illustrating the construction of information networks:

1. Collect data regarding the task under analysis, in this example through communication transcripts.
2. Use content analysis to identify key words in the data.
3. Take first information element and link (based on causal links within task) to other information elements. Repeat for each information element.
4. Use network statistics, or five links rule, to identify key information elements.

The information networks allow for the representation of all information used throughout the scenario (Walker, Stanton, Baber et al. 2010), including that which underpins decision-making (Stanton, Baber and Harris 2008) and enable the exploration of both systemic SA, individual SA and the manner in which SA develops over time phases (Walker, Stanton, Baber et al. 2010, Walker, Gibson et al. 2006, Stanton, Stewart et al. 2006). Within this research the focus is on illustrating the SA of the system as a whole, in line with the theory of DSA. In light of this information, networks have been created at the system level. The original networks are too large to be presented here; in order to focus analysis on the most salient

Table 3.6   Key information elements for the actual and ideal scenarios

Key Information Element	Actual	Ideal
Summarised ROE		
Army Ops Prep		
AWACS mission		
ATO		
Air Ops Prep		
F-15 mission		
Air Ops		
ACO		
Army Ops		
SITREP		
BH mission		
EN-ROUTE IFF mode		
EN-ROUTE radio freq.		
AWACS tracking		
F-15 report contact		
Briefing board		
MCC schedule		
Full ROE		

features of the network graph theory metrics, as used in the Social Network Analysis, they were employed to identify the key information elements in the actual and ideal networks. Key information elements are identified as those with a sociometric value of one standard deviation above the mean (Salmon, Stanton, Walker and Jenkins 2009, Houghton, Baber, McMaster et al. 2006). Table 3.6 above presents the key information elements (shaded) within the actual and ideal scenarios.

Examination of the key information elements reveals further differences between the actual and ideal scenario. The differences between the scenarios are illustrative of variance in information dissemination, for example within the actual scenario, *Briefing board* and *MCC Schedule* are not key elements. This is because within the actual scenario the information that was contained on the briefing board in the F-15 briefing room was not acknowledged by the F-15 pilots, specifically information concerning the Black Hawk flight. If the F-15 pilots had identified the briefing board information, the incident of fratricide could have been prevented, as in the ideal scenario. The same is true of knowledge of the *MCC Schedule*, again containing the Black Hawk flight information that might have alerted the F-15 pilots to friendly aircraft within the TAOR.

In this way examination of the information networks can reveal problems in information dissemination that may have contributed to the incident of fratricide, and may enable the development of appropriate preventions.

First presented by Stanton, Stewart et al. (2006), and continued by Walker, et al. (2009), is the argument that semantic networks, such as information networks, can be used to illustrate situation awareness from a systems perspective. The method enables an exploration of situation awareness elements within context, focusing on the relationships between elements (Stanton, Stewart et al. 2006, Walker, Stanton, Kazi et al. 2009). This information can be used to explore and understand the information needs of individuals and, moreover, how one might go about supporting those (Stanton, Stewart et al. 2006, Walker, Stanton, Baber et al. 2010). Information networks revealed the way in which actors within the scenario used the information they had. Specifically, it was revealed that a number of key information elements were not available in the actual scenario but were available in the ideal scenario. The ownership of key information elements such as the *MCC SITREP* would have been pivotal in the prevention of the fratricide incident had it been distributed fully in the actual scenario.

The presence, or availability, of information within a system does not guarantee that the information will be processed. As Boiney argues:

Information must also be attended to for SA to develop. (Boiney 2007: 2)

Activation of knowledge is very important and allows an illustration of what people were aware of, and what information they require (Stanton, Stewart et al. 2006, Walker, Gibson et al. 2006). According to Bolstad, Endsley and Cuevas:

Incoming data must not only be accessible, but filtered, analysed, and integrated to develop an overall situational picture to support team coordination and shared SA. (2009: 147)

Hourizi and Johnson (2003) discuss the notion that the presence of information is very different to the process of attending to that information, passing a large amount of information does not mean that the receiver will attend to all of that information and transfer it into a useful form. Hourizi and Johnson (2003) discuss three ways in which transfer of information may not equate to a greater level of situation awareness:

1. Information is available but not noticed.
2. Information is noticed but overlooked as full attention was not given,
3. Information is attended to but misunderstood, not interpreted in the appropriate manner.

EAST is able not only to illustrate the information within a system but also, through its information networks, to identify which information elements are activated and by whom. EAST enables the analyst to go even further and identify possible reasons for the activation of information elements through its exploration of schemata.

Schemata

Schemata are constructs which guide individuals' interaction with the world and affect the way in which individuals interpret that world (Neisser 1976). Langan-Fox, Wirth, Code and Langfield-Smith highlight a number of problems associated with schemata measurement, stating that an effective measure of such constructs has 'eluded researchers' (2001: 99). Klein and Hoffman also emphasise the complexity attached to exploring such constructs, stating that they cannot be 'observed directly in "pure" behaviour' (2008: 59). Despite the problems attached to schema measurement, Klein and Hoffman argue that 'mental representations can be inferred from empirical data' (2008: 64) such as that drawn from the information people utilise in task performance and agree with the notion put forward by Langan-Fox et al. emphasising the utility of using 'a network of nodes and links' such as 'concept maps, trees and semantic networks' (Langan-Fox et al. 2001: 100) to explore mental constructs.

The importance attached to 'faulty' internal mental constructs has been highlighted in multiple analysis of the scenario (Leveson 2002, 2004, Leveson, Allen and Storey 2002, Snook 2000). Leveson in particular provides an in-depth analysis of the 'inaccurate mental models' held by agents within the military system in particular of the F-15 pilots 'inaccurate model of the current airspace occupants' (2002, 2004).

The development of information networks allows for an exploration of the correct and the incorrect schemata held in the ideal and the actual systems. In previous applications of EAST analysis, schemata have been defined as those information elements that are persistent through time and are present throughout the phases of the scenario (Stanton, Salmon, Walker and Jenkins 2009a). This principle is used to guide the identification of schemata within the actual and ideal scenarios. The complete information networks for the scenario as a whole were supplemented with information networks developed for separate stages of the incident. The mission was broken down into 11 distinct time phases based on key events. On the following page, Table 3.8 and Table 3.9 contain a summary of the key information elements used in each phase of the actual and ideal scenarios; and Table 3.7 presents a key for the summary of key information elements, enabling an exploration of the key elements present in the actual and the ideal scenarios and how this presence differs between the two scenarios.

Table 3.7   Key for information element tables

Key	
Present in both scenarios	
Present in this scenario only	
Not present in this scenario	

Table 3.8   Actual scenario key information elements

Actual	8:22	9:21	9:27	9:54	10:12	10:15	10:20	10:21	10:22	10:24	10:25	10:28
Summarised ROE												
Army Ops Prep												
AWACS Mission												
Air Tasking Order												
Air Ops Prep												
F-15 Mission												
Air Ops												
ACO												
Army Ops												
SITREP												
BH Mission												
Enroute IFF mode												
Enroute radio freq.												
AWACS tracking												
F-15 contact report												

Table 3.9 Ideal scenario key information elements

Ideal	8:22	9:21	9:27	9:54	10:12	10:15	10:20	10:21	10:22	10:24	10:25	10:28
Full ROE												
Army Ops Prep												
AWACS Mission												
Air Tasking Order												
Air Ops Prep												
F-15 Mission												
SITREP												
Briefing Board												
MCC Schedule												
AWACS tracking												

Table 3.9 illustrates the development of situational awareness within the ideal scenario by presenting the key information elements active at the separate phases of the incident.

Within this scenario the information element of summarised ROE was present across all 11 phases of the mission, defining it as a prominent schema. The summarised ROE represents an incorrect mental template regarding the ROE for the mission. The F-15 pilots did not hold appropriate expectations regarding the manner in which they should react to a 'contact'. With regard to the ROE schemata, the full ROE specified that F-15 pilots must conduct a Visual Identification (VID) pass of the contacts to ensure that they were enemy targets. The F-15 pilot did conduct a VID pass of the two helicopters but at an excessive speed, altitude and distance. Failure to follow the ROE in this manner meant that the F-15 pilot was unable to gain an appropriate view of the helicopters. If an appropriate view had been gained then the pilot would have seen the American flags painted on the sides of the helicopters.

The table also highlights the long-standing presence of *BH mission* as a key information element, but only after the Black Hawks entered the TAOR at 09:21. This illustrates that the knowledge of the Black Hawk mission was not disseminated throughout the system before the Black Hawks entered the TAOR.

A core schema within this analysis is the ROE: despite all other actions if the F-15 pilots had held a correct knowledge of the ROE the incident of fratricide could have been prevented.

Table 3.10 below presents a summary of the information derived from the EAST analysis aligned with the factors of the F3 model.

Table 3.10   Summary of EAST analysis

		Actual	Ideal	
EAST	HTA	Linear chain of command, very separate tasks	Networked system undertaking tasks collaboratively	Coop. and Coord.
	HTA	43 extra bad tasks, 66 missing good tasks	66 extra good tasks, 43 missing bad tasks	
	CDA	77 team work tasks	98 team work tasks	Coord.
	CDA	1.93 overall coordination score	2.7 overall coordination score	
	SNA	Convoluted comms. structure	Comms. structure dominated by three key agents	Comms and Coop.
	SNA	Lower levels of sociometric status: Higher agencies need to be more prominent	Higher levels of sociometric status	
	SNA	Higher level of cooperation	Lower level of cooperation	
	IN	No correct schemata for all agents	Higher level of correct schemata	SA and Schemata
	IN	Lower level of key info elements	Higher level of key info elements	F3 Model

Table 3.11   Application of F3 model factors to erroneous actions

Error	F3 Model Factor
Failure to provide clear guidance leading to an unclear understanding of roles, specifically regarding helicopters	Coordination
Helicopters were not consistently regarded as part of Air operations or adequately integrated or monitored	Cooperation
Insufficent training on ROEs leading to a simplified understanding	Schemata
AWACS knowledge of Black Hawk helicopters was not passed to F-15 pilots	Communication
Black Hawk helicopter pilots were unaware of the correct radio frequency or IFF mode	Schemata
F-15 pilots misidentified the Black Hawk helicopters due to poor training and additional fuel tanks	SA
The Black Hawk helicopters had outdated radio equipment, preventing them from talking to F-15 pilots	Communication

Erroneous Actions

At a surface level the seven errors identified in previous analyses of the incident map onto the five factors of the F3 model, as illustrated above in Table 3.11.

However, as was discussed earlier, such superficial allocation of factors does not explore the causality behind these labels. In order to explore the reasons why breakdowns in these factors occurred, the method must identify and explore these breakdowns.

Through the HTA EAST was able to explore all tasks and actions that occurred within the accident scenario, including any erroneous actions. The method is able to explore each of the core seven errors common to the government report and previous analyses of the accident.

The creation of the HTAs allowed for a comparison of the tasks that occurred, or would have occurred, in the two scenarios. A number of core tasks that should not or did not occur in the accident scenario were identified. This allowed for the development of a set of initiating factors (confirmed by the government report as well as a number of other analyses of the same incident). In summary, there were 43 additional tasks that should not have occurred in the fratricide scenario and 66 tasks that should have occurred, but did not.

The presence of these findings in previous analyses bodes well for the sensitivity of EAST and its application to fratricide. An example of the crossover between the non-compatible tasks and the erroneous actions is task 1.3: compatible technology evolution, present in the ideal scenario HTA. This leads on to task 1.3.1, 1.3.2 and 1.3.3, which ensure that all OPC aircraft are fitted with the same radio equipment and can communicate with one another, ensuring that the Black Hawks and F-15s are able to use their radio equipment to communicate with one another. These tasks are not present within the actual HTA and due to this the Black Hawks and F-15s were not able to communicate with one another – a key error identified by the government report and previous analyses of the incident.

Relationships

Chapter 1 presented a SNA of the literature in order to explore the way in which the factors of the F3 model interact with one another. Within Chapter 2 the utility of this approach was reinforced by identifying positive and negative links in the 2003 Challenger II fratricide incident and using Social Network Analysis to model and explore these interactions. Within this section a similar analysis is presented for this incident of fratricide. The events that occurred within the mission were classified using the same coding scheme presented in Chapter 2.

Figure 3.10 presents an extract of the events and coding to illustrate the manner in which the coding was undertaken. The coding was conducted by the lead author in collaboration with an independent analyst. Both undertook an independent coding analysis of the scenario and then discussed the coding analysis until a definitive set of coding was produced.

Following the procedure outlined in Chapter 2, matrices were developed to illustrate the positive and negative links between factors and, from these, social networks were derived. Figure 3.11, on page 78, represents the model with a total figure for the number of negative links that occurred during the accident scenario. The thickness of the lines represents the quantitative value of each of the negative links. The dashed lines in the model represent links from the initial F3 model derived from the literature and presented in Chapter 1, which were not found within this case study.

Examination of this model and comparison to the model derived from the literature in Chapter 1, shown in Figure 3.12, illustrates a correspondence between the two. In both models the most prominent link is between communication and Situation Awareness.

Comparison of the two models also illustrates that the cooperation factor is not strongly linked to the other factors in the model. It is the author's opinion that cooperation, an attitudinal construct (Fiore, Salas et al. 2003, Wilson, Salas et al. 2007), is difficult to measure and assess. This explains the lack of literature linking this factor to other factors and also the lack of representation of cooperation within the event line. Additionally, the three links associated with cooperation are the weakest links in the model derived from the literature; they have the lowest quantitative value associated with them and therefore are the least prevalent. This means that within the fratricide example only the weakest links in the model were not captured, or not present, in this case.

With respect to the other links, although they are smaller in size in the model derived from the incident of fratricide, they are all reasonably comparable with one another, that is, they are of similar sizes to one another, which is true of the model derived from the literature as well.

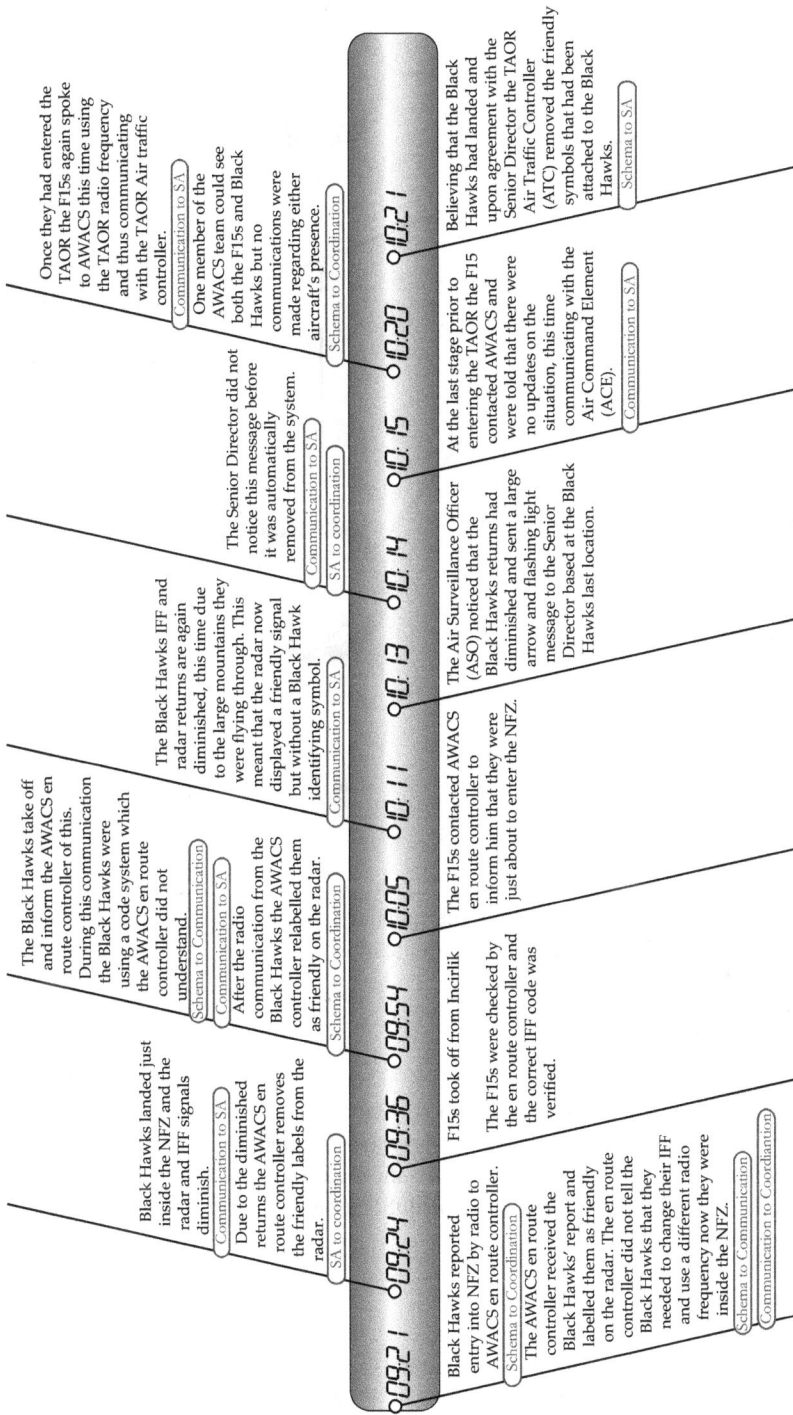

Figure 3.10 Timeline and coding

Timeline markers: 09.21 09.24 09.36 09.54 10.05 10.11 10.13 10.14 10.15 10.20 10.21

Black Hawks reported entry into NFZ by radio to AWACS en route controller.
[Schema to Coordination]

The AWACS en route controller received the Black Hawks' report and labelled them as friendly on the radar. The en route controller did not tell the Black Hawks that they needed to change their IFF and use a different radio frequency now they were inside the NFZ.
[Schema to Communication] [Communication to Coordination]

Black Hawks landed just inside the NFZ and the radar and IFF signals diminish.
[Communication to SA]

Due to the diminished returns the AWACS en route controller removes the friendly labels from the radar.
[SA to coordination]

The Black Hawks take off and inform the AWACS en route controller of this. During this communication the Black Hawks were using a code system which the AWACS en route controller did not understand.
[Schema to Communication]
[Communication to SA]

After the radio communication from the AWACS Black Hawks the AWACS en route controller relabelled them as friendly on the radar.
[Schema to Coordination]

The Black Hawks IFF and radar returns are again diminished, this time due to the large mountains they were flying through. This meant that the radar now displayed a friendly signal but without a Black Hawk identifying symbol.
[Communication to SA]

F15s took off from Incirlik

The F15s were checked by the en route controller and the correct IFF code was verified.
[Schema to SA]

The F15s contacted AWACS en route controller to inform him that they were just about to enter the NFZ.

The Air Surveillance Officer (ASO) noticed that the Black Hawks returns had diminished and sent a large arrow and flashing light message to the Senior Director based at the Black Hawks last location.

The Senior Director did not notice this message before it was automatically removed from the system.
[Communication to SA]
[SA to coordination]

At the last stage prior to entering the TAOR the F15 contacted AWACS and were told that there were no updates on the situation, this time communicating with the Air Command Element (ACE).
[Communication to SA]

Once they had entered the TAOR the F15s again spoke to AWACS this time using the TAOR radio frequency and thus communicating with the TAOR Air traffic controller.
[Communication to SA]

One member of the AWACS team could see both the F15s and Black Hawks but no communications were made regarding either aircraft's presence.
[Schema to Coordination]

Believing that the Black Hawks had landed and upon agreement with the Senior Director the TAOR Air Traffic Controller (ATC) removed the friendly symbols that had been attached to the Black Hawks.
[Schema to SA]

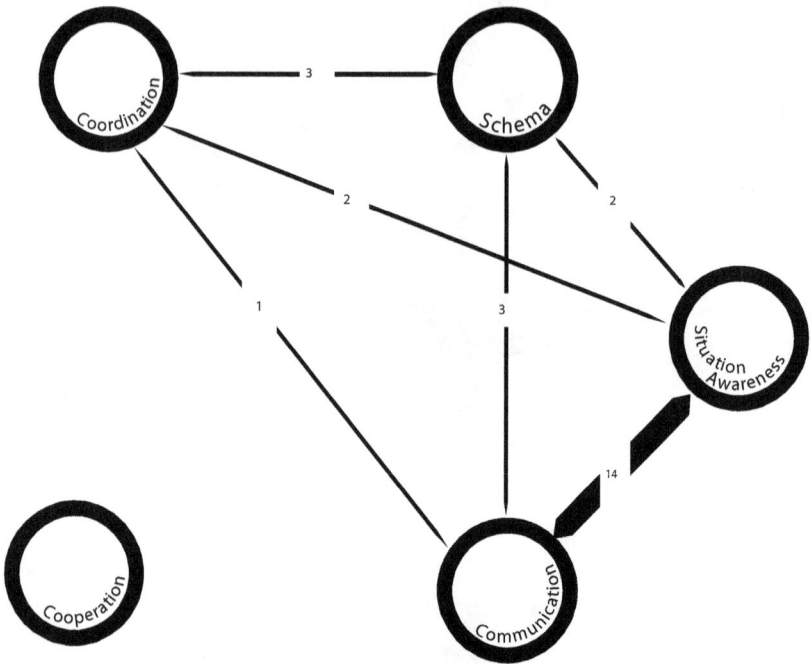

Figure 3.11   F3 model with quantitative values derived from the accident investigation report

System Components

Through its utilisation of SNA, EAST explores all system components involved in the accident scenario, as well as the manner in which they interact with one another as explored above, under Communication and Cooperation.

Guidance

Guidance about how to conduct the EAST method is available from multiple journal articles, book chapters and a book (Stanton, Baber and Harris 2008). The EAST methodology has been said to be easy to apply and the addition of software tools such as AGNA (Benta 2003), Leximancer (Leximancer 2009) and WESTT (Houghton, Baber, Cowton, and Stanton 2008) automate a large part of the analysis (Stanton, Baber and Harris 2008).

Local Rationality

Local rationality is explored through the HTA and the information networks utilised within the EAST methodology. The HTA presents the goals of both the

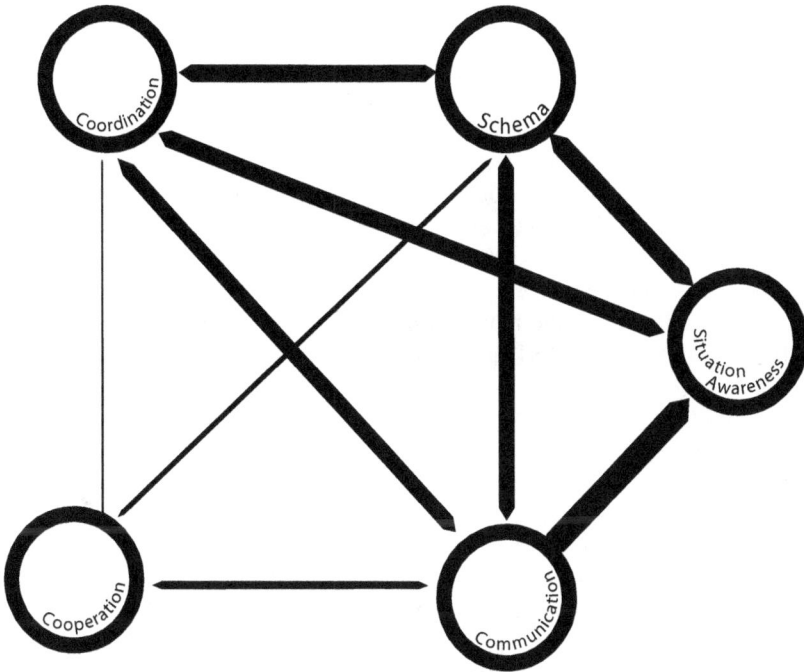

Figure 3.12   Literature-based F3 model

system and the individuals involved, and the information networks present the activated knowledge, Situation Awareness, both at a system and individual level.

Conclusion

The previous chapters have identified the need for a complex systemic method for the exploration of incidents of fratricide, and to provide validation for the F3 model. The EAST methodology was applied to an incident of fratricide and an ideal version of the scenario in order to ascertain its ability to analyse incidents of fratricide. EAST was able to meet all six of the criteria developed and enabled data to be inputted into each section of the model: Communication, Cooperation, Coordination, Schemata and Situation Awareness. The populated models were explored in order to understand the differences between the ideal and actual fratricide scenarios, highlighting the key causal factors.

EAST was able not only to provide a representation of all of the F3 model's causal factors but also able to provide quantitative measurements of these five factors. This enables statistical comparison of the causal factors between teams and between scenarios, providing a greater depth of exploration of the causal factors behind breakdowns, specifically for the comparison of the actual fratricide and ideal scenarios. Such quantitative metrics can also be used in future analyses to ascertain whether there are any similarities across incidents.

Furthermore, the analysis allows an exploration of the manner in which these causal factors interact with one another. The relationships between these factors and the manner in which they affect one another are the core focus of the F3 model.

From this comparison of the EAST results and coding, with the accident investigation results, it can be concluded that the factors, and links between the factors, included in the Famous Five of Fratricide model are representative of the events discussed by the accident investigation report for this incident of fratricide. This is a foundation for the model in explaining the causality of fratricide and shows the utility of the EAST method of assessing these causal factors and their interrelations.

In addition to addressing the key areas of the EAST method's applicability and providing construct validation of the F3 model, this analysis has raised a number of interesting points. One such point is that there may be metrics for the F3 model factors at which they are most efficient. If levels are too low, or too high, this may indicate a disposition for fratricide. This notion will be explored further through this book. The analysis also allowed for the identification of core differences between the scenarios; the fratricide scenario in particular was characterised by:

1. Lower levels of *cooperation* in goal completion and low levels of cohesive communication links between agents.
2. Lower levels of *coordination*.
3. Lower levels of *communication*, especially from higher systemic levels.
4. Inferior distribution of key information throughout the system resulting in poor *Situation Awareness*.
5. Greater prevalence of incorrect *schemata*.

This provides evidence for the applicability of both the EAST method and the F3 model; however, the research within this chapter explored a single case study and thus is couched at the initial phase of investigation. At this stage it must be pointed out that it is the authors' opinion that the event line derived from the accident report and other investigations does not provide a comprehensive account of all factors that had an effect on the accident scenario. For a more comprehensive analysis the analysts must experience the data first hand, to reduce interference or the influence of opinions of other analysts. Previous research into accident analysis has highlighted the limitations associated with exploring accidents based upon accident investigation reports (Griffin et al. 2010, Greitzer and Andrews 2008, 2009). Future work will involve first-hand experience of such accident scenarios in order to reduce these biases.

Chapter 4

It's Good to Talk: Exploration of the Communications Surrounding Shoot, No Shoot Decisions

Introduction

Chapter 3 demonstrated the applicability of the Event Analysis of Systemic Teamwork (EAST) method and coding as an appropriate methodology for the investigation of the causality surrounding incidents of fratricide. The method was utilised to explore an incident of fratricide and the resulting analysis highlighted the ability of EAST to meet all of the criteria put forward. The methodology aligns with the theoretical stance of this research and its multiple methods map onto each of the five factors of the F3 model, allowing explicit investigation of each factor. The addition of coding-based modelling enables illumination of the interaction between the factors.

Within this chapter the application of the EAST method to an incident of fratricide occurring within a military training scenario, compared to a non-fratricide version of the scenario, is presented. The performances of two teams were analysed: one in which an incident of fratricide occurred; and another in which the same mission was successfully completed, with only enemy targets engaged. The EAST outputs populated two prototypical models of fratricide and non-fratricide performance. The comparison revealed that the team involved in the incident of fratricide engaged in lower levels of communication, coordination and cooperation and held a lower level of appropriate schemata and accurate SA compared to the team that was not involved in an incident of fratricide. These factors, and their relationships, are predicted by the F3 model, and align with the research presented in the previous chapters.

Method

Simulators

The case study analysed within this chapter focuses on the military training of a Battle Group (a military force created to fight together, typically consisting of several different types of troops' (www.oxforddictionaries.com 2011)). The research involved undertaking multiple observations at a military training facility in which a series of high fidelity vehicle simulators were connected to create a virtual battle environment, enabling Battle Group level tank crew training. The friendly forces were 'played' by the training audience and the enemy forces were 'played' by trainers at the institution, serving Subject Matter Experts (SMEs).

Simulated environments provide a way in which to safely explore the decision-making processes undertaken by military teams and such simulated environments are often used in research (Orasanu 2005). Salas, Cooke and Rosen claim that the use of simulators represents an advance in research within the teamwork domain, as simulators:

> provide a valuable compromise between the complexity of the real world, which is an important influence on team performance and critical for establishing external valid results, and experimental control, which is necessary to establish internally valid results. (2008: 543)

Research into shoot, no-shoot decisions in other domains has successfully utilised simulators; for example Mitchell and Flin (2007) explored police shooting decisions using a high fidelity simulated environment. In addition to this, research specifically focused on fratricide has emphasised the benefits of simulated exercises to provide further knowledge surrounding incidents of fratricide (Gadsen and Outteridge 2006).

Scenario

The Battle Group analysed within this chapter consisted of B Company (14 Warriors – armoured personnel carriers), B Squadron (14 Challenger II Tanks – armoured tanks), F Squadron (13 Scimitars – armoured reconnaissance vehicles), as well as additional Recce (Reconnaissance), Engineer and Artillery troops. In total, over 50 vehicle simulators took part in the Battle Group training. The command structure of the Battle Group observed is presented in Figure 4.1.

The specific mission analysed within this chapter was a combined Battle Group quick attack, which lasted for 2 hours, 20 minutes. In this scenario the Battle Group was tasked with conducting a quick attack on a known enemy location. The plan was for a Reconnaissance troop to move up in order to identify a Forming Up Position (FUP; an area to gather prior to the attack) and a Fire Support Position (FSP; an area for a portion of troops to occupy in order to provide the attacking troops with defensive fire protection), with some support from a troop of armoured Recce (F Squadron). The mission performance and communications of all teams undertaking the training scenario were recorded and, post-scenario, two teams were chosen for in-depth analysis and comparison.

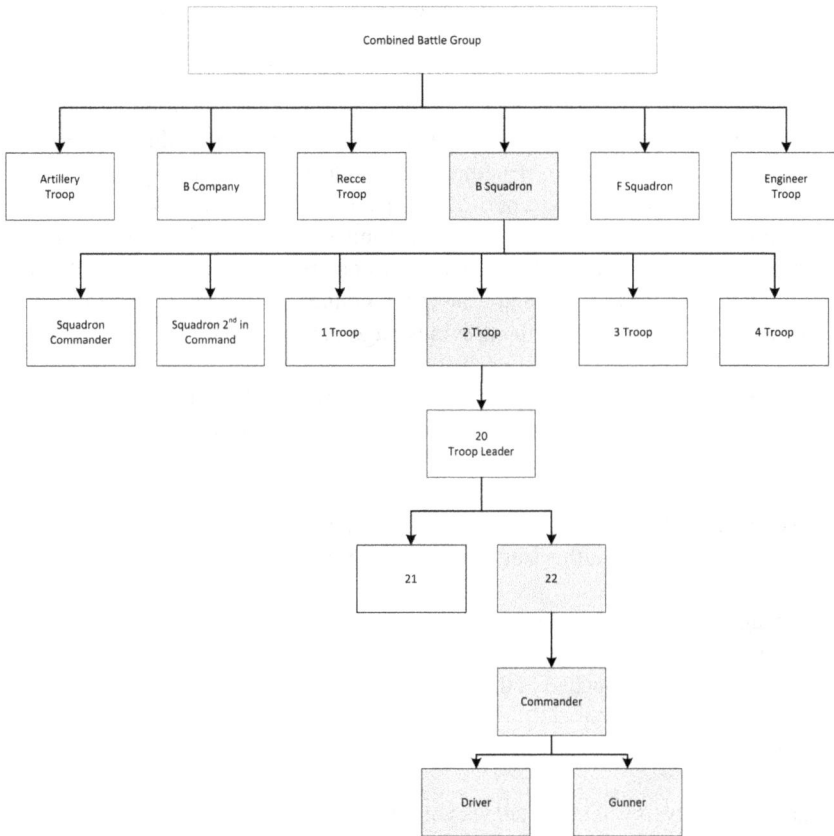

Figure 4.1   Command structure of Battle Group

Actual versus Ideal Performance

In Chapter 2 and Chapter 3 the utility of comparing ideal and actual performance has been emphasised. Although it is recognised that in such complex systems identification of less effective performance is difficult (Orasanu 2005), Subject Matter Experts (SMEs; trainers at the institution) were used to identify representative teams and an additional criterion was placed on the identification of a less effective team – they had to have engaged in an incident of friendly fire, during the training scenario, to be categorised as making less effective decisions.

During the training exercise an incident of fratricide was observed which occurred in a Challenger II tank crew. In order to ensure that the comparison between more and less effective performance was closely matched, the analysis aimed to explore the comparison with another Challenger II tank crew. The SMEs had identified a Challenger II tank crew whose performance was deemed to be representative of effective mission completion. In addition to this the comparison tank crew wad taken from the same vehicle type, the tank crew was also taken from the same troop, 2 Troop, and the same position under the troop commander. The utilisation of this tank crew in the comparison meant that the mission role, mission experience and team structure were as closely matched as possible.

Both teams analysed were part of B Squadron. B Squadron was made up of four troops of three tanks, as well as a commander and a second in command. This analysis focuses on one of these troops, 2 troop, specifically on two of its three tanks. Each tank crew contained a driver, a gunner and a tank commander. The shaded cells in Figure 4.1 on the previous page identify the tank crews compared within this chapter.

The two teams were chosen for analysis to represent tank crews who were able to effectively complete the mission, engaging enemy targets (subsequently referred to as the 'more effective' team), and a tank crew who completed the mission in a less effective manner, engaging a friendly reconnaissance vehicle (subsequently referred to as the 'less effective' team).

It must be stressed at this stage that the members of the Battle Group observed were in the initial stages of pre-deployment training. They return to the training facility again after an additional number of months training together before they are deployed. The actions they make cannot be judged against deployment standards and instead must be couched at the preliminary training level.

Data Collection

Data for the analysis was collected via voice recorders located in each tank simulator to record intra-tank communications, as well as recordings of all external tank communications (tank – tank communications, tank – HQ communications). The voice recorders used were Olympus VN-2100PC digital voice recorders. These recorders were linked directly to the radio headsets within the tank to provide a high-quality recording. In total over 1,600 communication acts were analysed. The exercise was observed from a set of control computer terminals which allowed for a detailed series of notes to be collected; providing an overview of all tank activity.

Data Reduction and Analysis

The voice recordings were transcribed and combined with the notes taken during observation, transcription of conversations with SMEs, and data automatically recorded by the computer system at the training institution, including weapons fired, distance travelled and so on. The data was then analysed using the EAST method. The outputs of EAST highlight differences between the more effective and less effective mission completion with respect to the five factors of the F3 model. Differences in the interactions between the factors of the two teams were also compared through the construction of a number of models: communication transcript coding enabled the development of a model representing the positive and negative links between the factors in both teams, creating four comparative models.

Inter-rater Reliability

In order to address the consistency of the implementation of the coding within this chapter, the inter-rater reliability of the coding was assessed for the CDA element of EAST, the communication coding and the coding-based population of the F3 model. The same procedure was followed for each inter-rater reliability study. A ten per cent selection of the HTA (for the CDA coding), or the communication transcript (for the communication coding and coding-based population of the F3 model), was provided to two independent analysts. The independent analysts were supplied with a set of coding rules to guide their rating. An explanation was also provided for each task on the HTA, and for each communication act on the communication transcript, to ensure that the raters could comprehend technical language and so forth. The raters were then asked to code the HTA or communication transcript. The results of this coding were compared to that of the original analyst and a percentage of agreement between the raters was derived in line with previous research (Crichton and Flin 2004, Baysari, McIntosh and Wilson 2008). Marques and McCall (2005) provide a review of inter-rater reliability, concluding that an acceptable level of agreement can generally be defined as 80 per cent agreement or above when reliability is calculated as the number of agreements divided by the total number of agreements and disagreements. In the relevant parts of this chapter the results of this inter-rater reliability will be presented.

EAST Findings

Overview

The more effective team began their mission with a clear brief and mission outline from the tank commander. This ensured that the team were aware of their goals, roles and what to expect throughout the mission. The more effective team then checked their position in relation to the tanks around them and ensured that this was correct. The team undertook the mission with high levels of communication, communicating relevant information to one another throughout; for example, any appropriate updates from the external radio networks were relayed by the commander to the rest of the team. The more effective team successfully completed their mission, moving to the correct locations at the appropriate times.

In the less effective team, no internal tank mission brief was given by the tank commander and information was less effectively communicated. The commander did not pay as much attention to the external radio networks and did not communicate all relevant information relayed on the external radio network to the team. The less effective team engaged the silhouette of a tank partially obscured by a tree line, believing it to be an enemy tank. Almost immediately after the engagement, the less effective team positively identified the target as a friendly reconnaissance vehicle, which confirmed that they had not gained positive identification before they engaged the target.

The next part of this chapter will present each of the F3 model's factors and explore the results that emerged from the EAST analysis, followed by a summary of the overall findings.

Coordination and Cooperation

Hierarchical Task Analysis (HTA)   A Hierarchical Task Analysis was used to describe each team's mission completion, using a hierarchy of goals, sub-goals and tasks. Every activity undertaken by the team was analysed in this manner, starting from the overall goal of the mission and working down to the individual tasks that occurred in order to achieve this goal (Annett 2005).

The HTA for each team was developed by studying the mission briefs, transcripts and observational notes. The culmination of these materials allowed for the key goals of the scenario to be identified. The overarching goal of 'Combined Battle Group quick attack' was identified from the materials as it was explicitly stated as the mission goal. This goal was then divided into five main sub-goals, which made up the overall goal of conducting the quick attack. These sub-goals represent the main phases of the mission:

*   Prepare for mission.
*   Move to objective.
*   Receive and transmit updates.
*   Identify targets.
*   Engage targets.

For every sub-goal, all of the actions that were conducted throughout the mission to meet this sub-goal were listed. At this stage the observable activities performed throughout the mission were assigned to the appropriate sub-goal and goal structure. In this manner each action performed within the scenario was represented in a hierarchical manner, illustrating the end goal of the action.

The complete HTA for the more effective team and less effective team are not presented in this chapter due to space constraints but are summarised by the high-level task networks presented in Figures 4.2 and 4.3. The task networks reveal five high-level goals that differ between the two teams, illustrated by the grey shading of goals present in the more effective team task network but not in the less effective team task network.

The differences in the high-level goals between the less effective and more effective team illustrate the different focus with which each approached the mission. The more effective team were aware that conducting *periodic situation*

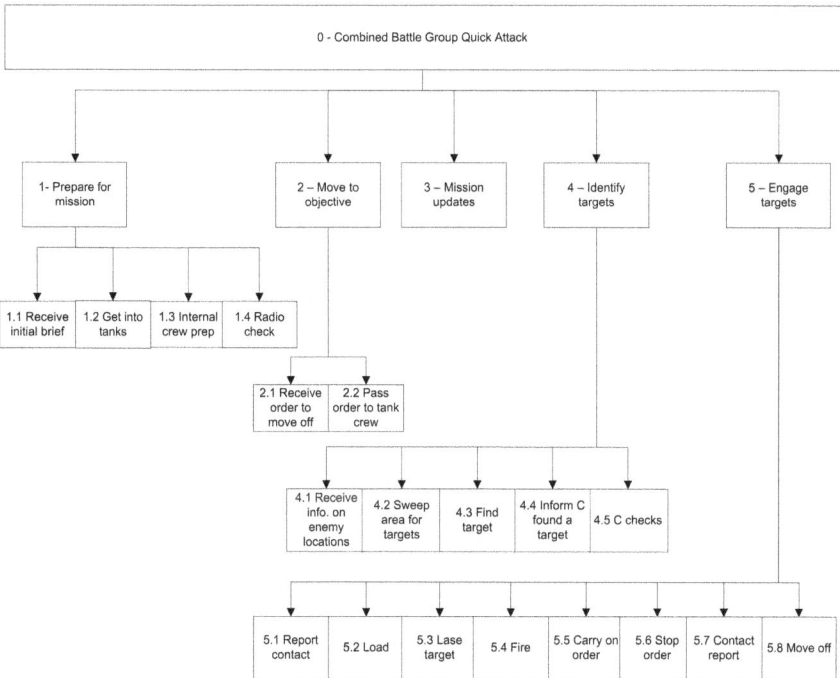

Figure 4.2   Less effective team task network

updates was a core goal needed to ensure that the tank crew could maintain accurate SA. This was not a core goal for the less effective team, indicating a lower focus on passing information within the less effective team. The notion of a lower emphasis on information transfer is strengthened by the presence of the *relay information to crew* goal in the more effective team but not in the less effective team. The additional goal differences of *Commander checks boundary lines*; *Commander checks COMBAT* (a computerised battle space visualisation tool); and *Commander checks IFF* (Identify Friend or Foe technology, which interrogates targets to determine if they are friendly) and *CID* markings (Combat Identification markings such as orange panels indicating a friendly vehicle), present in the more effective team but not in the less effective team, illustrate differences in the overall goal of *identify targets*. The more effective team aimed to complete the *identify target* goal by following each of the Standard Operating Procedures (SOPs) laid out. The less effective team, however, did not follow each of these procedures in their attempt to *identify targets*.

A more detailed analysis of the HTAs revealed a lower level of task steps within the less effective crew, 69 compared to 96 task steps in the more effective crew, as illustrated in Figure 4.4 on the following page.

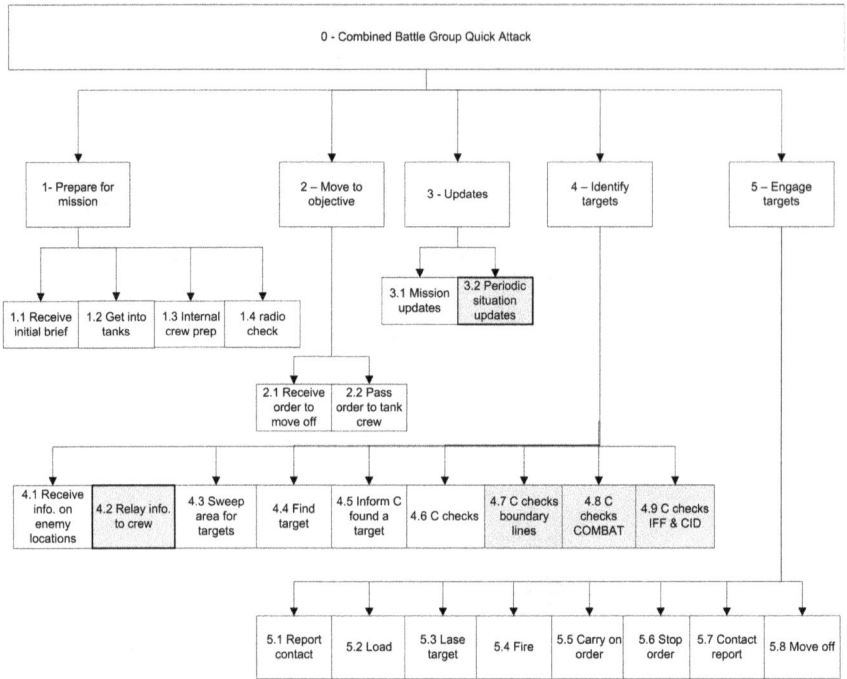

Figure 4.3   More effective team task network

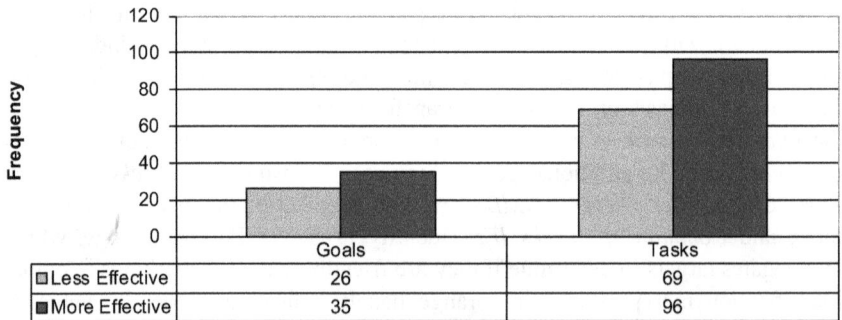

Figure 4.4   Comparison of goals and tasks

Detailed comparison of these tasks revealed that the less effective crew did not undertake a number of required tasks such as distributing external communications, including updated enemy locations, to the crew, failing to relay information regarding updates, intentions, expectations, plans and so forth.

The HTA results reveal that despite the confines of military orders, Rules Of Engagement and Standard Operating Procedures, the two teams performed the same mission in different ways.

Coordination

Coordination Demands Analysis (CDA) Using the two HTAs as inputs, Coordination Demands Analysis (Burke 2005) was carried out on the two teams' mission completion. The CDA allows for division and comparison of the tasks undertaken during mission completion that involved coordination between team members, teamwork tasks, and those which could be performed alone, individual tasks (Burke 2005). Tasks are extracted from the HTA and rated as either individual work tasks or teamwork tasks (Burke 2005). Teamwork tasks are then rated on a scale of 1–3 over six coordination dimensions to represent the degree of coordination in the team activities (Burke 2005).

The CDA undertaken within this chapter was subject to inter-rater reliability tests following the procedure outlined above under 'Method'. The results of the reliability testing revealed a mean agreement of 98 per cent with the initial CDA analysis carried out by the authors. Table 4.1 below presents a summary of the results of the CDA analysis for the less effective and more effective team.

Table 4.1   CDA analysis

	Less effective	More effective
Total task steps	216	181
Total individual work	56 (26%)	46 (25%)
Total teamwork	160 (74%)	135 (75%)
Mean Total Co-ordination	2.4 (80%)	2.9 (97%)

The results of the CDA analysis indicate that a greater number of coordinated tasks was carried out by the more effective team compared with the less effective team. Evaluation of the percentage of team tasks compared to the overall tasks undertaken by each team reveals little difference between the two teams, with both teams conducting a similar level of team tasks in relation to the total number of tasks they undertook, 74 per cent in the less effective team and 72 per cent in the more effective team.

The tasks carried out by the more effective team were correlated with higher coordination ratings (3 on a rating scale of 1–3) than those carried out by the less effective crew (1.98 on a rating scale of 1–3). This equates to a 100 per cent level of coordinated activity in the more effective team and a 66 per cent level of coordinated activity in the less effective team, in relation to teamwork tasks. From this it can be assumed that in this case study coordination affects team performance through the manner in which teams use it, rather than the quantity or frequency of team tasks undertaken.

CDA is normally used to illustrate the level of coordinated activity required within a scenario. SMEs are asked to rate each teamwork task in terms of the behaviour required to conduct it (Stanton, Salmon, Walker, Baber and Jenkins 2005). In the analysis presented here the teamwork required for each team to complete the mission would be the same; therefore the analysis focused instead upon the teamwork supplied by each team during the mission. As has been discussed previously there are numerous benefits associated with comparing more effective and less effective performance as opposed to the comparison of behaviour against a formal, prescriptive description (Orasanu 2005, Zsambok 1997).

Communication and Cooperation

Social Network Analysis (SNA)   Social Network Analysis enables the representation of the communication interactions that occurred during mission completion. Social network matrices were developed for both the more effective and the less effective teams' mission completion, containing all communications that occurred between all individuals during the mission. From the matrices, social network diagrams for the two teams were developed, as shown in Figure's 4.5 and 4.6.

The thickness of the arrows represents the frequency of the communication acts; the thicker the arrow the higher the frequency of communication between

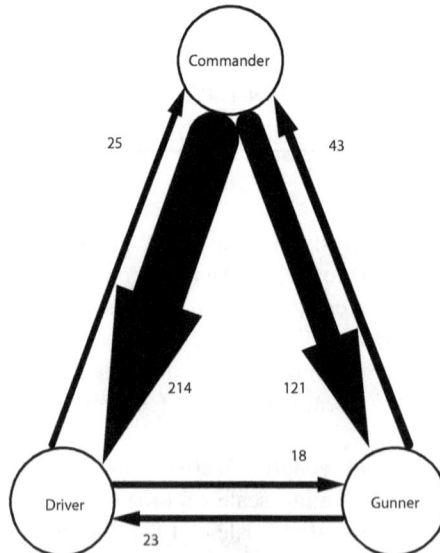

Figure 4.5   Less effective team social network

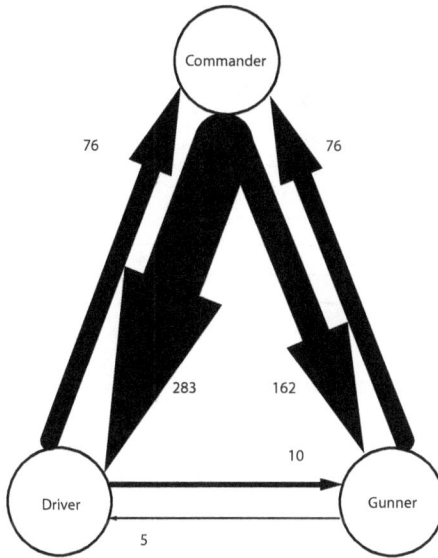

Figure 4.6   More effective team social network

the team members. The number next to each arrow represents the total number of communications acts between those two individuals. Examination of the social networks reveals differences between the social organisations in the two teams. In the more effective team the communication almost always involved the commander, with only 15 of 612 (2.5 per cent) communications not including him. Although the commander was still the dominant communicator in the less effective team, a higher number of communication acts did not involve him, 41 out of 444 (9.3 per cent). The results of the SNA illustrate a more hierarchical communication structure present within the more effective team. This is in line with research by Commungs and Cross (2003; cited in Katz, Lzaer, Arrow and Contractor 2005), which found that within small groups containing fewer than 12 members, a hierarchical communication structure was preferable to a non-hierarchical communication structure. Bowers et al. state that although there are disadvantages associated with hierarchical teams that 'these performance differences tend to disappear under high time pressure' (Bowers et al. 1992: 2392) where a hierarchical structure is shown to be advantageous.

Once the social network has been created, a number of statistics can be derived using graph theory metrics (Driskell and Mullen 2005). Five metrics were compared in this chapter: sociometric status; centrality; density; diameter and cohesion.

The greater the level of sociometric status an individual has, the greater the contribution that individual makes to the communication flow of the network, illustrating how busy the individual is in communication terms (Houghton, Baber,

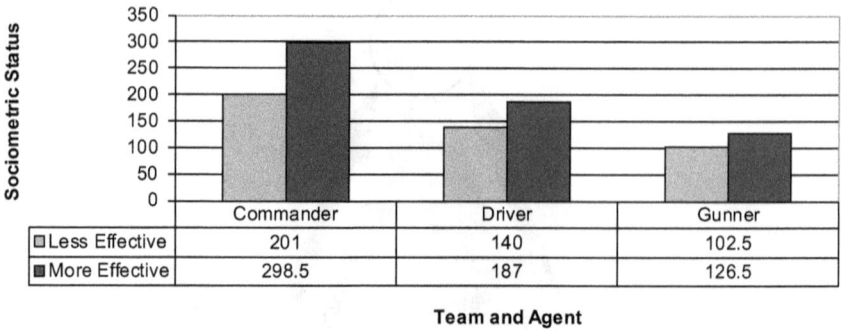

	Commander	Driver	Gunner
□ Less Effective	201	140	102.5
■ More Effective	298.5	187	126.5

Team and Agent

Figure 4.7   Sociometric status

McMaster et al. 2006). The individual with the highest level of sociometric status, the key individual, is the commander for both the less effective and the more effective teams; however, the level of sociometric status differs greatly between the two teams, as is shown above in Figure 4.7.

In the less effective team the commander has a lower sociometric status compared to the more effective team. From this it can be surmised that to have more effective team communications the commander must take a more prominent role in the communication. In the more effective team the sociometric status was higher for all members when compared to the sociometric status values for the less effective team. From the analysis it can be stated that the more effective team had a greater number of individuals making a greater contribution to the communication flow when compared to the less effective team. This finding is in line with the results of Chapter 3, where a greater level of communication from higher systemic levels was deemed beneficial. This conception is also supported by the wider literature, suggesting that a greater level of communication is valuable and, specifically, that leaders need to clearly articulate plans and intent (Cahillane et al. 2009, Bryant 2006, Shattuck and Woods 2000, Pigeau and McCann 2000).

There were no differences in the cohesion, diameter, density or centrality metrics between the more effective and less effective teams. Due to the small size of the team network it is not surprising that the cohesion, density and centrality metrics show no difference, these metrics represent aspects of the structure of the communication network, the number of connections between individuals within the network. In the small three-person team explored within this chapter there are only six possible communication links and these were utilised in both teams.

Communication

Communication Usage Diagram (CUD) A Communication Usage Diagram was developed for both the more effective and less effective teams. This involved identifying all communication acts that occurred during the mission (using the

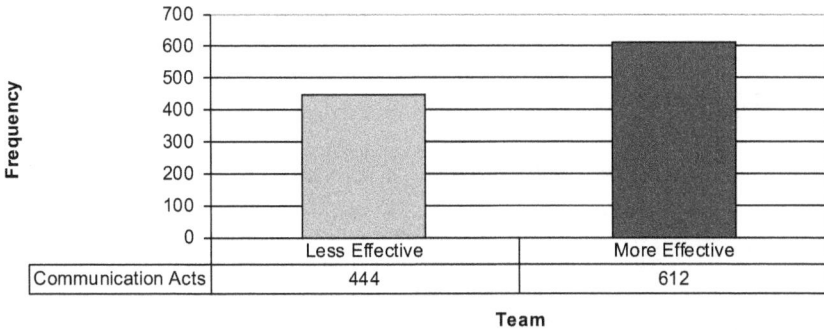

Figure 4.8   Summary of more effective and less effective team communications

communication transcript) and the means through which these occurred – in this case study entirely through the use of radios. A summary of the total communication acts in the teams' mission completion is shown above in Figure 4.8.

Although the CUD is normally used to critique the communication media employed (Stanton, Salmon, Walker and Jenkins 2005), within this case study all communication observed occurred through the use of radio networks. The CUD also provides the total number of communication acts that occurred during the mission: this total differed between the two teams. Communication within the more effective team was more frequent, with more communication acts than for the less effective team. Again, the findings of the CUD emphasise the notion that a higher frequency of communication acts is beneficial for effective decision-making and that a lower level may be detrimental.

Coding   In order to draw some more specific findings from the data, the communication acts were explored in more detail. This approach to performance assessment has been advocated by researchers such as Cooke and Gorman, who state that communication analysis allows for 'finer-grain holistic assessment of team cognition' (2006: 274) and suggest that a combination of flow and content analysis provide 'a window into the team-level cognitive processes underlying team cognition' (274).

The complete transcripts from the teams were coded into general communication types. The coding categories can be seen in Table 4.2, developed from the transcripts using a grounded theory approach (Glaser and Strauss 1967). The transcripts for each team were divided into meaningful chunks of information and categorised into a series of high-level categories such as *transmit*. The transcripts were then further separated into more specific categories based upon the explicit detailed content, the type of information contained within them. Codes were developed in order to ensure that all communications were fittingly categorised. Unfortunately, due to the quality of the voice recordings, six communication acts

Table 4.2   Coding scheme

Code	Rule for coding
Transmit information	A communication that passes explicit information and updates Situation Awareness
	Example 'on the right hand side'
Transmit direction	A communication that informs the driver of where to go, which direction to take, any communication involving the word 'stick' as tanks use a left stick and a right stick to steer
	Example 'come left stick, on'
Transmit order	A communication that includes a specific order
	Example 'Fire!'
	(all orders relating to direction should be in the transmit direction category)
Transmit review	A communication provides information on the current situation
	Example 'that is the rest of the Battle Group there'
Transmit	Any other communication which transmits but does not fit into one of the above categories
	Example 'good' or 'sorry'
Non	Any non-mission related communication; for example, talking about that evening's dinner
	Example 'My key just fell down the tiniest little gap in the world'
Request information	A communication in which an explicit request for information is made
	Example 'Can you still see him'
Request confirmation	A communication in which a specific request is made for a confirmation that the communication has been heard, for example, a repetition of the communication, or asking for a repetition of the communication
	Example 'Here?'
Acknowledge	A communication that provides a person with an acknowledgement, for example that an order/ piece of information has been heard
	Example 'Yeah' or 'Right'
Radio	A communication involving information regarding the radio network or radio headsets
	Example 'Can you here me?' or 'radio check'
Grid reference	A communication involving a grid reference – any six digit number
	Example '650148'
Transmit expectation/ plan	A communication containing information about the mission plan or about expectancies of what the mission will involve
	Example 'they will be moving in a minute'

could not be clearly deciphered and therefore could not be coded. The coding of the communication transcripts was subjected to inter-rater reliability testing following the procedure outlined above under 'Methods'. The results of this analysis revealed a mean inter-rater reliability agreement of 82 per cent.

An example of an extract from the communication transcript taken from the less effective team is presented below:

> Don't stop keep going, just slow it down a bit, okay slow it down.

This extract illustrates the commander directing the driver and was coded as '*transmit direction*'. The frequency of these communication types present in the communication transcripts of the two teams is shown below in Table 4.3.

Table 4.3   Coded communication acts

	Less effective n (%)	More effective n (%)
Transmit information	43 (9.7%)	113 (18%)
Request information	27 (6%)	91 (15%)
Direction	127 (28.6%)	135 (22%)
Order	13 (3%)	23 (4%)
Transmit exp/plan/review	3 (0.7%)	53 (8.6%)
Transmit	22 (5%)	49 (8%)
Non-mission related	176 (39.6%)	86 (14%)
Request confirmation	5 (1%)	13 (2%)
Radio	7 (1.6%)	13 (2%)
Transmit confirmation	2 (0.5%)	12 (2%)
Grid reference	6 (1.3%)	4 (0.7%)
Cant code	2 (0.5%)	4 (0.7%)

Table 4.3 reveals that the less effective team communicated a much lower level of communications relating to *transmit information* than the more effective team as a percentage of, and in relation to, the overall communication acts within each team. Researchers such as Hirokawa and Johnston (1989) and Hollnagel (2007) emphasise that without communication of information team members cannot be expected to make effective decisions. These findings provide further support for the assumption that lower levels of communication within a team may act as a causal factor for fratricide.

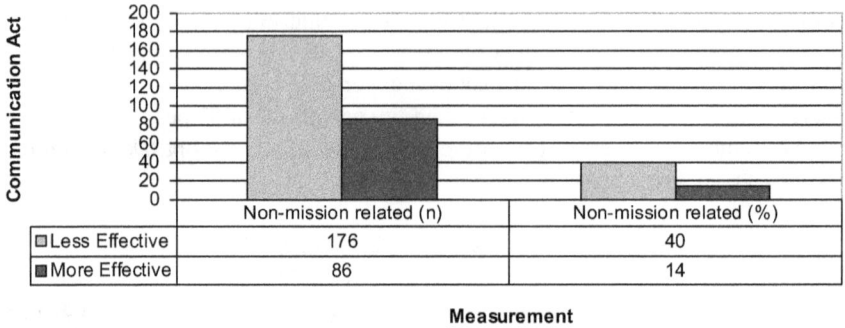

Figure 4.9   Non-mission related communication acts

As is illustrated by Table 4.3, after the *transmit information* category the largest difference between the two teams is in the category labelled *non-mission related*. This category refers to the number of *non-mission related* communication acts that occurred within the mission. Again this difference is based upon the percentage of *non-mission related* communication acts in relation to the overall level of communication acts within each team. Figure 4.9 above focuses on this difference, supporting research by Hutchins, Hocevar and Kemple (1999), which indicated that poorly performing teams fail to keep non-essential communication to a minimum.

Of all communication acts within the less effective team, 39.6 per cent were non-mission related. This is more than double the percentage of communication acts coded as non-mission related in the more effective team, again in relation to the percentage of non-mission related communications compared to the overall number of communication acts within each team. According to Manning (1991) a low level of non-mission related communications may be beneficial for effective decision-making, in order to build and maintain levels of cooperation within the team and a feeling of comradeship. Once these communication acts reach a certain level, however, they may begin to distract crew members from their mission and thus become a factor in fratricide incidents. Non-relevant communications may be detrimental in two key ways: firstly, non-essential levels of communication can increase the level of cognitive resources required for communication and can distract individuals from other tasks (Dismukes, Loukopoulos and Jobe 2001); and, secondly, they can block important communications on both the internal and external radio nets.

Moore, Schermerhorn, Oonk and Morrison (2003) emphasise the importance of relevant information transfer in military teams, hypothesising that there are four fundamental types of inappropriate information transfer:

1. Producing information irrelevant to team member needs.
2. Not producing information relevant to team member needs.

3. Consuming irrelevant information.
4. Not consuming relevant information.

Moore et al. (2003) argue that these inappropriate information transfers can degrade SA as they hamper the ability to access required information. The ability to access the required information can be impeded by such information not being attended to and to irrelevant information being attended to.

Situation Awareness

Information networks and Concept maps   A number of the analyses presented above have highlighted the benefits attached to a greater level of appropriate, mission-relevant communications. Here, the analysis explores whether a greater level of communication is indicative of a greater level of SA.

In Chapter 3 the use of information networks for exploring SA was discussed (Walker, Stanton, Baber et al. 2010, Walker, Gibson et al. 2006, Stanton, Stewart et al. 2006, Salmon, Stanton, Walker and Jenkins 2009). The contextual analysis associated with the development of information networks gives rise to a conflict between the drive for reliability and consistency, and the detailed insight provided by such contextual analysis. The research approach taken throughout this book has been to explore phenomena within their context, for example, naturalistic case study research and grounded theory analysis (Gillham 2000, Numagami 1998, Smith, W. and Dowell 2000, Salmon, Stanton, Walker, Baber, Jenkins and McMaster 2008, Farrington-Darby et al. 2006, Stanton, Rafferty et al. 2010, Walsham 1995, Darke, Shanks and Broadbent 1998, Glaser and Strauss 1967). A continuation of this contextual analysis therefore seems fitting. In order to improve the reliability of the data, this contextual analysis was supplemented with a computer software tool, Leximancer (Leximancer 2009), which is able to divide the data in a stable and reliable manner (Smith and Humphreys 2006). In this way both a subjective analysis and a more controlled analysis of the transcripts was undertaken.

Within this chapter, analysis of the communication transcripts and their ability to explore SA was conducted through the development of information networks and through the use of Leximancer, the contextual analysis software tool mentioned above. Other researchers have used similar software-based methods to explore information elements and their interconnections. Crichton, Flin and McGeorge (2005) outlined a technique to elicit information from on-scene commanders in nuclear emergencies. The method involved compiling a matrix of information which was analysed for its relatedness, that is, the relationships between information elements. The results of their analysis were visualised using a clustering technique to illustrate the conceptual relationships between the information elements – based on a frequency of co-occurrence. Walker, Stanton and Salmon (2011) utilised the Leximancer software to provide an exploration of the differences in mental representations of car drivers and motorcyclists for the same roads. The following part of this provides a discussion and comparison of information networks and Leximancer Concept maps.

As is discussed in Chapter 3, information networks provide an exploration of: information elements (the total number of pieces of information transferred during the mission); key information elements (the most frequently transferred or most central information elements as defined by graph theory metrics, such as sociometric status); and the manner in which these information elements are linked (information network connections and graph theory metrics such as density).

Leximancer provides a set of concept seeds (the total number of pieces of information transferred during the mission); a set of concepts (the most central – frequently occurring – information elements); and the manner in which these concepts are linked (concept map connections, levels of relevance and connectivity for each concept and a representation of the most likely pathways between concepts). In addition to this, Leximancer provides a thematic analysis of the data showing the manner in which the concepts identified are divided into higher-level thematic categories.

Walker, Stanton and Salmon (2011) and Smith and Humprehys (2006) provide descriptions of the stages involved in the Leximancer analysis as summarised here:

1. The analyst should cleanse and format transcript data (check spelling and consistency of term use).
2. The analyst should modify the default Leximancer settings for the analysis of transcript data.
3. Leximancer then identifies all concepts present within the transcript data.
4. Based on co-occurrence and relationships between concepts Leximancer identifies the most important concepts from the transcript.
5. The analyst can then explore and edit the emergent concept seeds, ensuring that concepts are not replicated and are relevant.
6. Leximancer uses the relationships identified earlier to develop a concept map clustering and linking elements appropriately.
7. Leximancer produces concept and thematic summaries allowing an overview of the most important elements and their interactions with one another; pathways between elements can be explored both graphically and by exploring transcript extracts presented by the software.

Concept maps, such as those derived from Leximancer, have an established history in the representation of information. Crandall, Klein and Hoffman (2006) provide an in-depth description of concept maps, exploring their inception in the work of Novak while he attempted to explore and track the manner in which his students learned. Crandall, Klein and Hoffman discuss the utility of such methods in eliciting and representing knowledge exploring the methods basis in Ausubel's theory of learning, which posits that 'meaningful learning takes place by the assimilation of new concepts and propositions into existing concepts and propositional frameworks held by the learner' (2006: 43). Crandall,

Klein and Hoffman (2006) describe concept maps in a very similar vein to the description provided earlier of information networks – they consist of a set of nodes (representing concepts) which are linked to one another in meaningful ways through the application of lines. Walker, Stanton and Salmon (2011) illustrate the use of Leximancer-derived concept maps and information networks to explore SA and schemata. They state that the work of researchers such as Salmon, Stanton and Walker (2009) and Stanton, Salmon, Walker and Jenkins (2009b) has 'anchored' the methods to Neisser's Perceptual Cycle and to schema theory, allowing for a theoretically valid exploration of the 'mental representation of a situation' (Walker, Stanton and Salmon 2011: 5). Stanton, Salmon, Walker and Jenkins (2009b) utilise networks to illustrate system level SA. The networks use element nodes to illustrate information required to complete the goals of the system and conceptual links to represent causal links that occurred between the elements during task performance (Stanton, Salmon, Walker and Jenkins 2009b), positing that communication transcripts can be used to represent phenotype schemata.

Stanton, Salmon, Walker and Jenkins (2009b) discuss the manner in which the information networks can be decomposed to describe the SA of individuals within the system. They state that exploration of the interaction between individuals in the system allows for an investigation of emergent phenotype schemata. They continue, arguing that exploration of multiple phenotype schemata allows inferences to be made regarding the underlying genotype schemata. Indeed, it has been argued that Leximancer enables the exploration of emergent behaviour and unknown relationships through its ability to 'discover implicit, indirect relationships between concepts' (Smith and Humphrey 2006: 264).

A comparison of a sample of the key elements, identified through the development of information networks and through the use of Leximancer, revealed a percentage agreement of over 80 per cent, representing a high level of correlation between the analyses. In light of this, Leximancer has been used to explore SA and schemata throughout the remainder of this book because of the additional benefits associated with the method in terms of reliability, repeatability and the exploration of themes within the data.

The software explores the transcript data both in terms of conceptual and relational links (see Leximancer 2009). The relationships between concepts are shown in a number of ways:

1. The colours of the theme circles represent how connected the themes are using heat mapping: the more red a theme is, the more connected it is (Leximancer 2009) (although this aspect of Leximancer is not explored within this book).
2. The distance between concepts represents their conceptual context; concepts that are positioned close together frequently occur with similar other concepts (Leximancer 2009).

3. The percentage relevance score represents 'the number of occurrences of the concept as a proportion of the most frequent concept'; it shows how central the concept is in the network (Leximancer 2009: 73).
4. The connectivity score shows how prominent the theme is and how connected the concepts within each theme are (Leximancer 2009).

In addition to these illustrations, Leximancer also allows for more detailed exploration of the links between concepts. The map illustrates the most likely path between any two concepts in the network, allowing for an exploration of indirect relationships (Leximancer 2009).

The graphical illustrations of the concept maps (referred to as 'information networks' from here on) in both teams, derived from Leximancer, are presented below in Figure 4.10 and Figure 4.11.

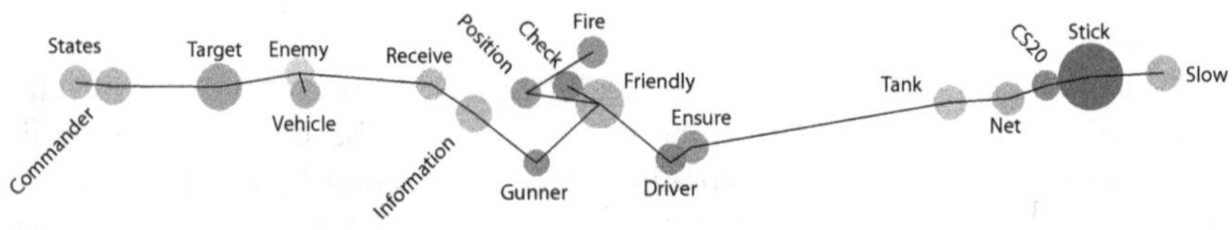

Figure 4.10   Less effective team information network

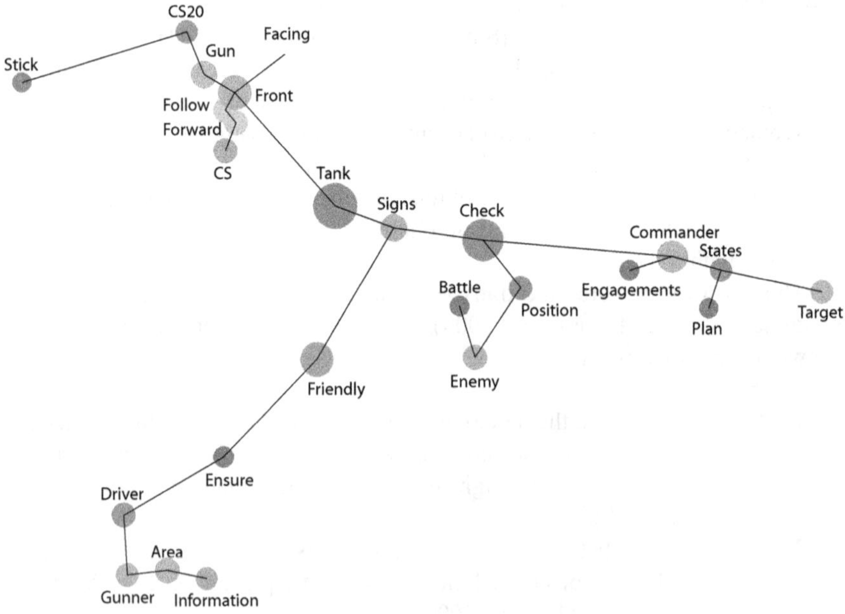

Figure 4.11   More effective team information network

From a visual examination, the elements within the more effective team appear to be organised in a more connective manner, with elements linking to more than one other element, whereas in the less effective team the connectivity appears to be linear, with elements linking into only one other element. The linear connectivity illustrated within the less effective team would equate to a greater distance between elements within the network as a whole, that is, a lower level of paths connecting elements to one another. To explore the relationships further, the elements and their associated relevance are presented below in Table 4.4. The shaded cells represent elements present in both the more effective and the less effective teams.

Table 4.4 **Elements and relevance for the more effective and the less effective teams**

Less effective			More effective		
Element	Count	Relevance	Element	Count	Relevance
target	13	100%	tank	16	100%
stick	12	92%	commander	15	94%
hold	12	92%	check	14	88%
information	7	54%	target	13	81%
commander	6	46%	front	10	62%
friendly	6	46%	enemy	10	62%
ensure	6	46%	friendly	9	56%
states	5	38%	information	9	56%
tank	5	38%	follow	7	44%
net	5	38%	gunner	7	44%
fire	5	38%	C/S 20	7	44%
enemy	4	31%	engagement	7	44%
receive	4	31%	gun	6	38%
slow	4	31%	call sign	6	38%
driver	4	31%	position	6	38%
gunner	4	31%	area	6	38%
vehicle	3	23%	driver	6	38%
position	3	23%	states	6	38%
check	3	23%	facing	6	38%
C/S 20	3	23%	ensure	6	38%
			plan	6	38%
			forward	5	31%
			signs	5	31%
			stick	5	31%
			battle	4	25%

A mean of 59 per cent (58 per cent of the more effective elements and 60 per cent of the less effective elements) of the elements are present in both the more effective and the less effective teams, illustrating that despite the two teams having undertaking the same scenario, the information they used differed greatly, with 40 per cent of the information differing between the two teams. Examination of the elements reveals the differences between the information present within the less effective and the more effective teams, but, more importantly, the table reveals the importance of these elements within the network and their relationships to one another. As is discussed throughout this book, it is not the information that is important but how it is interpreted, understood and integrated.

Table 4.4 reveals numerous differences in the relevance associated with the elements in the less effective and the more effective teams. The relevance metric refers to the individual elements and in order to explore the connectivity of the network as a whole, graph theory metrics were calculated from the Leximancer-derived information networks. Walker, Stanton, Salmon, Jenkins et al. (2010) illustrate the utility of the application of graph theory metrics to information networks, arguing that the metrics allow 'emergent properties of the participants' knowledge base, both in terms of interrelations and elements to be distilled and expressed in simple numeric terms' (2010: 479), providing a guide to the visual complexity of information networks. Five metrics were calculated in this analysis: sociometric status; centrality; density; cohesion and diameter. The results of this analysis are presented below in Table 4.5.

The results reveal that the more effective team information network contained a higher mean level of centrality, and a lower diameter value. The additional metrics were comparable between the two teams.

Centrality is defined as a metric representing how central each node is within the network the prominence of a node within a network (Walker, Stanton and Salmon 2011). From the results of this analysis it could be hypothesised that the more effective team contained a higher number of prominent information elements. A higher level of prominence suggests that the information in the more effective team was more appropriate to the task at hand, signifying that the more effective team may have been more capable of distinguishing appropriate from inappropriate information. The distinction between appropriate and inappropriate

Table 4.5   Information network metrics

	Less effective	More effective
Density	0.05	0.04
Mean Sociometric Status	0.1	0.08
Mean Centrality	13	19
Diameter	14	11

information has received a great deal of attention in the Human Factors domain (Bryant 2006, Boiney 2007, Moffat 2003). Boiney hypothesises that:

> human attention, rather than information, has become the critical resource for situation awareness and decision making in many command and control environments. (2007: 2)

The lower diameter value in the more effective team indicates that the paths between information elements were shorter, compared to the paths between information elements in the less effective team. Walker, Stanton and Salmon, in their exploration of individual cognition, suggest that as 'average lengths between semantic concepts decrease … suggested a more integrated mental representation' (2011: 883). They argue that mental representations can be argued to be more integrated as there is 'more direct access, without intervening concepts on the pathways across the network' (2011: 884). Applying this thinking to the cognition of a system (as analysed here) would suggest that information received by the system is less distributed in that elements have shorter pathways to one another. From this it could be hypothesised that in collocated team situations as analysed here, the more effective team, with a lower diameter value, had a more integrated information network than the less effective team, with a higher diameter value. The manner in which this information is integrated and interpreted can be explored further through thematic analysis.

Schemata

Thematic analysis   The thematic analysis derived from Leximancer enables an exploration of the prominent schemata present within the more effective and less effective teams. The analysis clusters elements together to allow for an exploration of the genotype pattern underlying them, the schemata. Each element is represented within a higher-level theme used to illustrate the underlying focus of the information element. The themes are drawn from the communication transcript based upon the co-occurrence of elements and extracts are presented to illustrate the manner in which these themes are drawn out, examples of which are presented below. Figure 4.12 and Figure 4.13 represent illustrations of the thematic analysis for the more effective and the less effective teams.

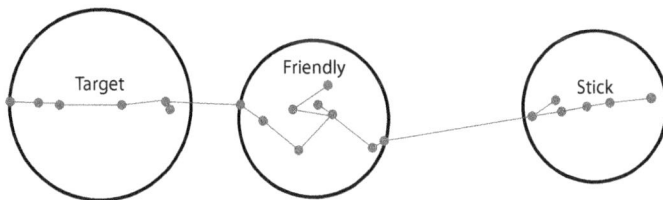

Figure 4.12   Less effective team themes

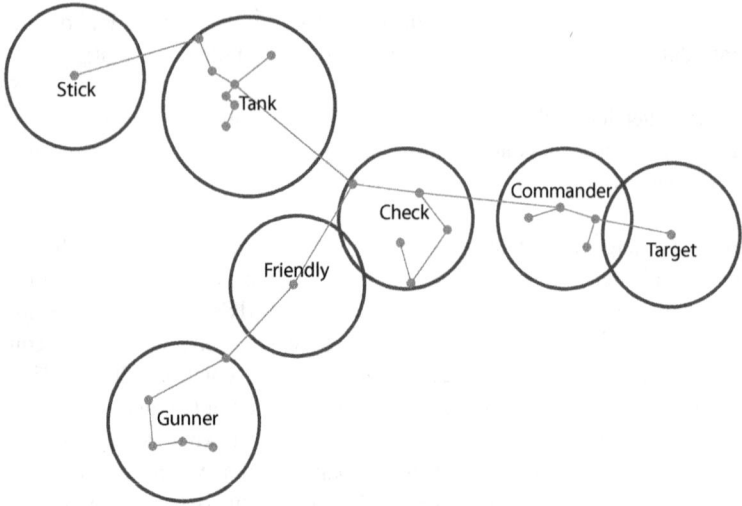

Figure 4.13   More effective team themes

The two figures illustrate the differences between the systemic schemata of the more effective and the less effective teams. The less effective team SA can be characterised as centred on three core themes, whereas the more effective team holds seven core themes which underlie its SA. Table 4.6 below provides greater insight into the thematic differences between the more effective and the less effective teams. The shaded cells represent themes present in both the more effective and the less effective teams.

Counter-intuitively, in light of the engagement by friendly troops, the *friendly* theme has a greater level of connectivity within the less effective team: 88 per

Table 4.6   Thematic analysis of the more effective and the less effective team

Less effective		More effective	
Theme	Connectivity	Theme	Connectivity
stick	100%	tank	100%
friendly	88%	check	54%
target	67%	gunner	38%
		Commander	34%
		friendly	17%
		target	9%
		stick	4%

cent, compared with 17 per cent in the more effective team. However, examination of the thematic analysis reveals that the theme *tank* has a relevance of 100 per cent within the more effective team, and this theme contains information solely discussing friendly tanks and the team's own tank in the more effective team. However, in the less effective team the theme of *friendly* is related to *fire*, which is inappropriate and represents the act of fratricide that occurred and was discussed within the transcript.

Within Leximancer a summary of the most relevant transcript extracts for each theme are presented. Of the four most relevant extracts for the less effective team, two were related to the incident of fratricide. In contrast, all of the most relevant extracts from the more effective team are related to ensuring that the location of friendly tanks is confirmed:

> Did you hear that – there's friendly dismounted troops where we are going now so just take it easy

> Can you see the friendly dismounts?

> Yeah, That's them in the trench there

The exploration of the thematic analysis reveals that information within the less effective team was incorrectly integrated as a result of the fratricide incident; but what incorrectly integrated or interpreted information led to the incident of fratricide? Perhaps the theme of *check*, present within the more effective scenario and highly relevant, 54 per cent, which is not present within the less effective thematic analysis or element analysis, can illustrate this point. The *check* theme incorporates elements such as *call signs, enemy* and *position*; these are important parts of the mission that are not represented in the thematic summary for the less effective team. From this it could be suggested that the less effective team failed to correctly interpret the requirements imposed upon them to maintain an awareness of friendly call sign positions and to ensure that enemy targets were not, in fact, friendly call signs.

The prominence of the theme *stick* within the less effective scenario (100 per cent relevance) compared with the more effective scenario (4 per cent relevance) serves to further highlight why the incident of fratricide may have occurred. Inside the tank, the driver uses two 'sticks' to steer the vehicle: when he pushes down the left stick the tank moves right; and when he pushes down the right stick the tank moves left. The difference in connectivity of the theme *stick* illustrates that within the less effective team the commander provided a low level of information to his team, instead concentrating on giving his driver a high level of basic instructions:

> Pick up left stick, hard left stick, right there

> Pick up left stick, left stick, hard there …

Come right stick on, right stick on,

Hard right stick, hold there

Reverse

Right stick

Hard /on

Left stick, hard left stick on

Left stick on, right stick on, stop

okay right stick, on, reverse,

Right stick, on, right stick, on, and hold it there, stop, good, that's us

On the other hand, in the more effective team a greater level of information relating to higher-level goals was passed to the driver, and the driver was left to direct himself, freeing the commander of this role:

Stay with C/S 21 we are going to the FUP. Try and stay in low ground as much as we can.

The commander's instructions could be classified as action-based within the less effective team, and effects-based within the more effective team. From this perspective the commander in the more effective team had a greater level of cognitive resources to attend to the mission scenario, as he was not responsible for providing a high level of detailed guidance to the driver, as was the case in the less effective team.

A contextual analysis of the two teams' transcripts reveals the manner in which information could be incorrectly interpreted and integrated in the less effective team. Many incorrect schemata led to the incorrect belief that the friendly reconnaissance vehicle engaged was enemy. In line with the terminology utilised by Norman (1981) these schemata can be divided into four 'parent' schemata, which have many 'child' schemata:

1. The less effective team's *pre-mission* preparation schema was incorrect. The less effective team did not check the *direction* of the tank, check that the tank was in the correct *formation*, or conduct an adequate *internal brief*. The lack of the formation check led to a lack of knowledge regarding other friendly-force location. This deficit in friendly-force location knowledge

was heightened by the lack of the internal brief which should have provided correct schemata on both enemy and friendly pictures, including locations, plans and numbers. In the more effective team the *pre-mission* schema was correct and as a consequence the more effective team did check the *direction* of the tank, did check its *formation*, and did conduct an *internal brief*. As a result of these actions the more effective team had a greater awareness of friendly-force locations and plans, as well as enemy locations and numbers.

2. The less effective team's *update* schema was incorrect: very few situation updates were passed to the rest of the team by the tank commander during the mission. This in turn created issues with both the *enemy* and *friendly* schemata, along with all their 'child' schemata. In the more effective team the *update* schema was appropriate and the tank commander frequently passed updates regarding his expectations about *enemy* forces and information regarding *friendly* forces.

3. The less effective team held an incorrect schema with regard to monitoring the external *radio networks*. These networks were not being monitored consistently and this led to a lack of information being available to the less effective team. In turn the *enemy* and *friendly* schemata were not appropriately updated. In the more effective team the *radio networks* schema was appropriate and the team effectively monitored the *radio networks* to ensure that they gathered all appropriate information and updates on *enemy* and *friendly* forces.

4. The *engage* schema held by the less effective team was incorrect. Specifically, the team believed that they could engage without following the correct identification *procedures*. Additionally, they appear to have held an incorrect schema regarding the visual *identification* of *enemy vehicles*. The more effective team appear to have a correct *engage* schema in addition to identification *procedures identification of enemy vehicles* schemata, illustrated by their ability to engage an enemy target correctly.

In summary, it is suggested that the less effective team members may have held an incorrect schema that there were no friendly call signs in their vicinity. The initial briefing they received before the mission began specified that all friendly call signs would move to the Forming Up Position except for 2 Troop, which would provide fire support for the Operational Commander as he brought up the rear of the formation. During the mission the less effective team did not receive any information to update this schema and therefore it was maintained. Although updates were sent over the radio networks, an incorrect schema regarding the monitoring of the radio networks meant that this information was not picked up by the less effective team. This incorrect schema regarding the radio networks also meant that they did not receive information regarding mission plan updates and friendly call sign locations. Additionally, the information that the commander did receive was not disseminated to the crew; he gave few updates through the mission and, indeed, did not give an initial briefing before the mission began. This meant that the crew did not have the key information needed to trigger correct schemata for the situation. This inappropriate schema meant that when the less effective team encountered a camouflaged vehicle they assumed it to be enemy and this led to the engagement of a friendly vehicle. In contrast, the more effective team were aware of friendly and enemy locations and did not engage a friendly vehicle.

The analysis of the networks developed through the use of the Leximancer software allowed for an illustration of the differences in SA and schemata, in relation to both content and structure, between the more effective and less effective teams. It appears that the more effective team's SA was more integrated and developed than that of the less effective team. This meant that the more effective team was able to interpret information they received throughout the scenario appropriately, whereas the less effective team was less capable of interpreting the world around them correctly.

Summary

Table 4.7 below represents a summary of the core findings derived from the EAST analysis detailed above.

Table 4.7   Summary of EAST analysis

		Less effective	More effective	
	HTA	Lower level of required tasks	Higher level of required tasks	
	HTA	Lower level of tasks – 95	Higher level of tasks – 131	
	CDA	Lower level of team tasks – 51	Higher level of team tasks – 69	Coord
	CDA	Lower overall coordination score 1.9	Higher overall coordination score 3	
	CUD	Lower level of communication acts – 444	Higher level of communication acts – 612	Comms
EAST	CUD	Higher level of non-mission comms – 39%	Lower level of non-mission comms – 14%	F3 Model
	CUD	Less hierarchical – 9% not through commander	More hierarchical – 2% not through commander	
	SNA	Lower mean Sociometric status – 148. Lower for all agents	Higher mean Sociometric status – 204. Higher for all agents	Comms and Coop
	IN	Information is not always correctly interpreted	Information is correctly interpreted	SA and Schemata
	IN	Mean Centrality is lower – 13	Mean centrality is higher – 19	
	IN	Diameter is higher – 15, information is less connected	Diameter is lower – 11, information is more connected	

The EAST analysis identified lower levels of coordination and cooperation in the less effective team. Lower levels of relevant communication acts were also found in this team, along with higher levels of non-relevant communication acts. The less effective team had a lower level of centrality, and a lower level of connectivity between, information elements suggestive of a less well developed SA and a lower ability to interpret information correctly.

Interaction of Factors

The EAST analysis has enabled the exploration of the factors affecting fratricide occurrence in two tank teams. In order to explore the interactions between these factors, that is, the systemic nature of their behaviour, further analysis is needed. The links between the factors were identified by stepping through the two teams' mission communication transcripts and using coding to identify the links in the model, as well as identifying breakdowns in these links. The procedure followed in this coding was the same as that outlined in Chapter 2 and Chapter 3, and the coding scheme for the links used was presented in Chapter 2. The population of the models was validated following the inter-rater reliability procedure outlined above under 'Methods'. The results of the analysis revealed a percentage agreement of over 80 per cent, illustrating that the coding of the communication transcripts was applied in a reliable and repeatable manner. Two examples of the way in which these links were identified can be seen below:

> Guys there's friendly dismounted troops where we are going now so just take it easy.

This first extract is taken from the more effective team's transcript and was labelled as representing the link *Communication – Situation Awareness*. The commander's effective communication to the team of an update he had received enabled the team to develop correct Situation Awareness.

> Um 12, is that 02 … and um 0 I don't know who that one is over there.

This second extract is taken from the less effective team's transcript and was labelled as representing a breakdown in the link *Coordination – Situation Awareness*. This extract is the gunner's response after the commander has asked what the call sign (C/S: 'the combination of identifying letters, letters and numbers, or words assigned to an operator, office, activity, vehicle, or station for use in communication (as in the address of a message sent by radio)' (Merriam Webster 2011)) of the tank next to them was. Due to a lack of coordination between the friendly call signs involved in the mission, the less effective team had poor SA regarding the identity of the tanks around them. In this way the mission transcripts enabled a representation of the positive links between factors within the model

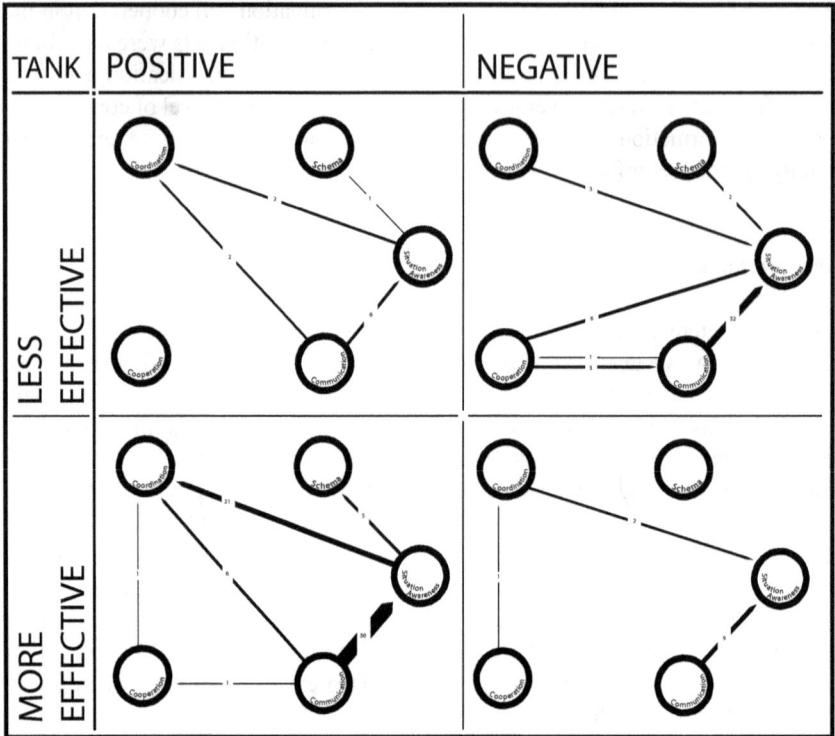

Figure 4.14 **Comparison of the positive and negative links for the less effective and the more effective teams**

to be derived, as well as identifying breakdowns, or negative links, between the factors in the model.

The populated F3 models for the less effective and the more effective teams are presented above. Figure 4.14 compares positive links (such as the above example of *Communication – Situation Awareness*) between factors and negative links (such as the above example of *Coordination – Situation Awareness*).

A graphical exploration of the models illustrates a greater number and frequency of links within the more effective positive model compared to the less effective positive model, and a greater number and frequency of links in the less effective negative model compared to the more effective negative model. Table 4.8 and Table 4.9 provide a summary of the frequency and presence of links in each of the models.

The tables illustrate that the less effective team only had four positive links and that each of these links held a low frequency – the thinner the lines, the fewer times this link occurred during the mission. This analysis reveals that the more effective team model had additional positive links and that these links were much higher in

Table 4.8   Frequency of positive and negative links for the more effective and the less effective teams

	Positive	Negative	Difference
Less Effective	11	47	-36
More Effective	84	12	72
Total	95	59	

Table 4.9   Presence of positive and negative links for the more effective and the less effective teams

	Positive	Negative	Difference
Less Effective	4	6	-2
More Effective	6	3	3
Total	10	9	1

frequency than in the less effective team, with much thicker lines. Comparing the two teams' positive links, it can be seen that Situation Awareness and its link to Communication was a prominent link in the more effective team, but this link was minimal in the less effective team. It is also clear that the link between Situation Awareness and Coordination was important in the more effective team and again nominal in the less effective team. From this it may be concluded that key aspects of effective decision-making in the more effective team – Situation Awareness and its links to Communication and Coordination – were barely present in the less effective team. The model also allows identification of the role of Cooperation in the more effective team – with its links to Communication and Coordination. In contrast, links to or from cooperation were absent from the less effective team.

The negative links discovered in the less effective team further highlight the importance of the link between Communication and Situation Awareness; this was a negative link numerous times within the less effective team. There were also an additional five negative links within the less effective team mission: the links between Situation Awareness and Coordination; Schemata and Situation Awareness; Cooperation and Situation Awareness; Cooperation and Communication; as well as Communication to Cooperation, were all negative. Again, the lack of Cooperation in the less effective team was highlighted. In contrast to this there were only three negative links in the more effective team's model. These negative links were between Communication and Situation Awareness; Cooperation and Coordination; and Coordination and Situation Awareness. It is interesting to note that both teams had negative links between Communication and Situation Awareness, but the frequency of this link was much higher in the less effective team than in the more effective team. Both teams also had negative links between Coordination and Situation Awareness, although the frequency of this link was not large for either team.

In summary, the link between Situation Awareness and Communication seems to be the most dramatic for the teams: in the more effective team, positive links between these two factors were higher, and breakdowns far lower. In the less effective team positive links were lower and negative links far higher. From this it can be argued that the link between Situation Awareness and Communication is a key factor in fratricide.

Summary of Results

There are three core, directly observable, findings derived from the research presented in this chapter:

1. The less effective team held an inappropriate schemata regarding the location of friendly tanks.
2. The less effective team received a poor level of information dissemination from the tank commander.
3. Communications in the less effective team were further negated by external communication-masking as a result of internal, non-mission relevant, communications.

These findings are supported by the fratricide and wider teamwork and safety literature presented within Chapter 1. Appropriate schemata are required in order to ensure that individuals correctly interpret and interact with the world around them (Neisser 1976, Norman 1991). The schemata held by individuals impacts on the manner in which they interact with the world; inappropriate schemata can prevent information from being attended to due to 'confirmatory bias' (Famewo, Matthews and Lamoureux 2007, Dean and Handley 2006, Greitzer and Andrews 2009) and can cause information to be interpreted incorrectly (Neisser 1976, Norman 1981). Good levels of information dissemination are required for effective team performance (Flin, Slaven and Stewart 1996, Svensson and Andersson 2006, Siegel and Federman 1973) and specifically in order to prevent incidents of fratricide (Jamieson and Wang 2007, Wilson, Salas, Priest and Andrews 2007, US Congress 1993), both to aid in the development of appropriate SA (Stout et al.,1999, Svensson and Andersson 2006) and activation of appropriate schemata (Flin, Slaven and Stewart 1996). Effective communication not only involves adequate information dissemination but it also requires that communication acts are appropriately attended to; thus non-mission relevant communications should be minimised as they can be detrimental due to their ability to distract and cognitively overload team members (Dismukes, Loukopoulos and Jobe 2001).

Conclusion

To conclude, the application of EAST to two teams' mission performance allowed data to be inputted into each node of the F3 model. The analysis reinforced the importance of Coordination, Cooperation, Communication, Situation Awareness and appropriate Schemata to effective team performance, with higher levels of these factors correlating with more effective team performance, offering empirical validation of the F3 model.

Coding allowed for an exploration of the links between the nodes. The results revealed that the link between Communication and Situation Awareness was consistently the most frequently negative link in a less effective team model. A new negative link between Cooperation and Situation Awareness was also identified during the tank crew study; this, along with the identification of other negative Cooperation links, verified the importance of Cooperation breakdowns.

The research emphasised the importance attached to communication in the fratricide literature, revealing that teams that communicate more on mission relevant information make better shoot, no-shoot decisions. This is because communication within a team is indispensable for good teamwork; it triggers appropriate schemata, informs SA and can encourage both cooperation and coordination. The more effective team communicated more and communicated more mission-relevant information.

Chapter 5

The Communication-masking Effect: Why it's Not Always Good to Talk

Introduction

Overview

Chapter 4 explored an incident of fratricide within a traditional tank crew-training environment. Modern warfare is consistently moving away from tank battles and towards joint force missions (Pirnie et al. 2005). The importance attached to joint force missions has been highlighted by the Ministry of Defence, which states that for future operations greater levels of 'air/land/maritime integration at far lower (systemic) levels will be required' and 'operations within any one environment will become increasingly dependent on cross-component capabilities' (Ministry of Defence Defence Concepts and Doctrine Centre 2008: 14). Previous research into fratricide has also emphasised the importance of exploring joint force missions such as Close Air Support tasks (Zobarich, Bruyn-Martin and Lamoureux 2009, Mistry et al. 2009). This chapter explores the way in which military teams make decisions within a Joint Fires mission training exercise. To investigate the decision-making process, more effective and less effective decisions were compared. The chapter begins with a discussion of the importance of effective decisions within the Joint Fires domain and the factors that have accentuated this need in modern times. The scenario under analysis is then described, along with the methodology utilised. The results of the EAST analysis are then summarised, allowing key insights to be drawn. Finally, a series of data-filled F3 models is explored and conclusions on the decision-making process are theorised.

Background to Joint Fires

This chapter explores the way in which military personnel make decisions within the context of air and land force integration. A number of training scenarios were observed in which the British Army and the Royal Air Force (RAF) worked together in pre-deployment training. The main aim of this training was to ensure that both the Army and the RAF understood how best to utilise one another's assets, specifically training them to work together on Joint Fires missions. Joint Fires missions require both forces to work collaboratively in order to engage a target; normally this will involve the Army calling in the RAF for Close Air Support (CAS) – to engage a target close to their position.

Modern warfare highlights the need for joint operations between the Army and the Air Force in numerous ways. Pirnie et al. (2005) explored the issue and stated that warfare has changed; the 'Cold War' style warfare of two large opposing forces has been replaced with asymmetric warfare, where enemies are far smaller targets, such as a single antagonist mingling in a crowd of civilians. Zobarich, Bruyn-Martin and Lamoureux also emphasise the changing nature of warfare, stating that 'targets are more often than not individuals or small structures rather than formations of mechanised infantry or large military HQ' (2009: 228). Pirnie et al. (2005) state that these smaller targets require greater precision weapons, which technological advances have provided and, as a consequence of this, people have become increasingly removed from the visual identification process. Decisions within the military domain are already complicated by factors such as time pressure, ambiguous problems, unstable conditions and severe consequences (Drillings and Serfaty 1997); the additional constraints imposed on decision-making by not actually being able to see a target highlight the importance of understanding this decision-making process in the hope of improving the effectiveness of decision-making in this context.

This chapter explores two examples of air–ground integration: in one a battery (a small group of artillery) effectively conducted the Joint Fires mission (called 'the more effective team'), where all targets were neutralised; in the other a battery conducted the Joint Fires mission less effectively (called 'the less effective team'), where not all targets were neutralised, a higher level of engagement from the enemy took place and an incident of fratricide occurred.

Chapter 4 explored an incident of fratricide at a low systemic level, a small, three-man tank crew. This chapter explores a much higher systemic level, the Joint Fires Cell (JFC). The Joint Fires Cell interacts with multiple assets, each with their own distinct role and task, during mission performance. Exploring fratricide within this environment provides an interesting addition, and comparison, to the results of the previous chapter.

Communication and the Communication-masking Effect

Chapter 4 revealed the importance held by the level of communication occurring within a mission. In this chapter the notion of communication is further explored, including the concept of communication overload – the inability of people to process vast amounts of information and the dire consequences excessive communication can have with respect to 'masking' relevant communications, due to both an inability to identify relevant communications and an inability to process all communications.

The issue of communication overload has been raised previously in fratricide research. Bolstad, Endsley and Cuevas argue that the technological advances of the modern world 'have resulted in a huge increase in systems, displays and technologies, particularly in complex operational environments such as the military' (2009: 148). They argue that such systems elongate and complicate the procedure of developing SA, as individuals must sort through a host of information inputs while also judging the reliability of the information feeds. They suggest that the dilemma of the modern world is no longer in gaining more information, but rather in:

> finding, within the volumes of data available, those precise bits of information that are needed to make an informed, reasoned decision. A widening gap exists between the tons of data being produced and disseminated, and the individual's ability to find the right, disparate bits and process them together to arrive at the actual information sought. (Bolstad et al. 2009: 148)

Barnet (2009) and Hinsz and Wallace (2009) contend that there is a point at which individuals can longer process all of the information they are exposed to, and that:

> when inappropriate or overwhelming information is provided, inaccurate identifications are made, and incidents of friendly fire may result. (Hinsz and Wallace 2009: 202)

Research by Moffat (2003) discusses the 'complexity penalty', stating that additional communication connections can lead to information overload and an inability to process information. Supporting this assertion is work by Maule (2010), which discusses the limitations of human decision-making. Maule cites the work of Miller (1956) and his discussion of the limit to the amount of information people can process. The work of researchers such as Reason (1990) and Woods et al. (1994) continue this emphasis, highlighting the inability of humans to process large amounts of information, contending that poor decisions are the result of activating, integrating or focusing on inappropriate information over relevant information. Reason discusses the problem of 'bounded rationality', referencing Simon (1975), who defines it:

> [T]he capacity of the human mind for formulating and solving complex problems is very small compared with the size of the problems whose solution is required for objectively rational behaviour in the real world. (Reason 1990: 198)

Fiore, Salas, Cuevas and Bowers (2003) emphasise the importance of communication efficiency in teams, especially within environments in which team members are distributed. They cite work by Hess et al. (2000), which states that in distributed teams the need for efficient information management increases due to the risk of overload. They argue that distributed teams have higher cognitive demands, that distribution itself 'may place additional demands on team members' (Fiore, Salas et al. 2003: 354). The idea that distribution requires additional resources is emphasised by numerous researchers such as Paris, Salas and

Cannon-Bowers (2000) and Boiney (2007), who all posit that distribution imposes supplementary strain and increases the resources required to maintain effective communication and coordination.

The supposition that lower levels of communication in certain circumstances are beneficial is supported by research such as that by Urban, Bowers, Monday and Morgan (1995), who identified that effective teams were more able to communicate efficiently only the appropriate information. Urban, Bowers et al. (1995) theorise that this is due to effective teams having appropriate schemata about which information is needed by whom in the team. As is stated by Gorman, Cooke and Winner, within military teams it is vital that:

> the right information is communicated to the right team member at the right time, and this involves team coordination. (Gorman, Cooke and Winner 2006: 1320).

Stanton, Stewart et al. (2006), Dismukes, Loukopoulos and Jobe (2001), Moore et al. (2003) and Hutchins, Hocevar and Kemple (1999) all highlight the detrimental aspects of excessive communication, suggesting that additional communications can be irrelevant and cause unnecessary cognitive workload. In a study analysing the communications occurring within a military brigade-level system, Walker, Stanton, Jenkins et al. (2009, 2010) found that a large amount of data transfer, utilising a large level of bandwidth, was used to transfer a comparatively small level of usable information. The importance of efficient communication and information transfer are emphasised within the military domain where bandwidth is habitually limited (Walker, Stanton, Jenkins et al. 2009, 2010).

Specifically exploring military commanders information processing, Bryant (2006) discusses the inability of the human mind to keep up with the increases in data presentation that have occurred with technological developments. Bryant concludes that 'the problem facing commanders is that access to data does not necessarily translate into useful information' (Bryant 2006: 186). Bryant also discusses the negative impact of presenting additional information to commanders, stating that higher levels of information could reduce a commander's understanding, causing him to 'struggle to process a deluge of raw data' (186). Research into the military domain by Bolstad, Endsley and Cuevas has also highlighted the notion that:

> the problem in these environments is no longer a lack of information, but finding, within the volumes of data available, those precise bits of information that are needed to make an informed, reasoned decision. (2009: 148)

Hourizi and Johnson (2003) continue to discuss the complex relationship between information and SA, stating that presenting more information will ensure that adequate SA can be developed, yet it can have a detrimental impact on SA development in that it could distract an individual's attention away from important information.

The reasons behind the positive effects of efficient communication are explored by Stanton, Stewart et al. (2006) who argue that, from the DSA perspective, additional communications can be detrimental to the development of SA in three main ways: adding additional, irrelevant, information and thus cognitive demands; delaying relevant information transfers; and shadowing the emphasis of relevant information. This is in line with the research presented here, which found that efficient communication of key information elements was correlated with effective performance.

Within this chapter an empirical evaluation of the information utilised in more effective and less effective shoot, no shoot decisions is provided in order to further explore the problem of information overload within incidents of fratricide.

Method

The core aim of the research presented by this book is to validate the F3 model using a variety of case studies, and this chapter represents one case study. The focus of this book is on the detailed analysis of each case study presented, rather than the exploration of numerous case studies. This approach is adopted because decision-making scenarios, specifically shoot, no-shoot decisions, are highly complex and as such require multifaceted analysis to develop an appropriate understanding of them. Reducing this analysis would lead to oversimplification and possibly result in inaccurate conclusions being drawn from the work.

Scenario

The observations analysed in this chapter took place in an aircraft simulator training facility developed to enable pre-deployment Close Air Support training. The training facility contains simulated and synthetic forces demonstrating allies, neutrals and enemy forces. Additionally, the facility contains Fire Planning Cell areas, Fire Support Team tents and Joint Fires Cell tents enabling Joint Fires mission training.

Operation Shura was the specific mission observed. Each Battery was tasked with attending a Shura – a meeting of village elders – alongside identifying and destroying an IED (Improvised Explosive Device) factory in the local vicinity. It must be stressed at this stage that the batteries observed were in the initial stages of pre-deployment training. They return to the training facility again after an additional number of months' training together before they are deployed. The actions they make cannot be judged against deployment standards and instead must be couched at the preliminary training level.

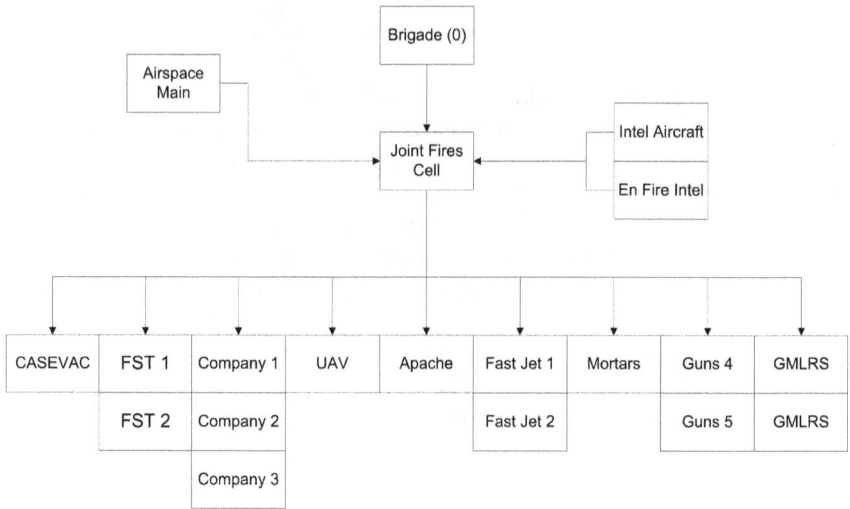

Figure 5.1   Command structure of the Battery

The system observed consisted of numerous assets, as is illustrated above in Figure 5.1, which gives an example of the initial command structure of the system for the more effective and the less effective teams.

Initially, the Joint Fires Cell controlled all fire and ISTAR (Intelligence, Surveillance, Target Acquisition, and Reconnaissance) assets – the fast jets and helicopters – Apache and CASEVAC (Casualty Evacuation helicopter), Unmanned Aerial Vehicle (UAV), guns, GPS Missile Launcher Rocket System (GMLRS) and mortars. During the mission the Joint Fire Cell was able to allocate control of these assets to the Fire Support Teams and control could even be allocated to the Company level. The JFC were controlled by Brigade throughout and were also under the over-watch of Airspace Main (which controls the airspace for a larger area). Additionally, the Joint Fires Cell received intelligence from an Intelligence Aircraft as well as from an Enemy Fire Intelligence asset; however, these assets were controlled at a higher level and, as with Airspace Main, acted for a larger area.

Data Collection

Data for the analysis was collected automatically by the computer system at the training centre. The voice communications between all elements of the system (a total of 12 radio networks) were then transcribed for analysis. In addition to this, the exercise was observed from the Fire Support Teams' tents, the Joint Fires Cells tent, as well as from a set of exercise control computer terminals which allowed for an overview of all activity. Notes taken during observation of the mission were then combined with transcriptions of all radio communications and conversations with Subject Matter Experts (SMEs) involved in the exercise in order to construct an accurate account of both batteries' mission performance.

Data Reduction and Analysis

In line with Chapter 3 and Chapter 4, a combination of the EAST methodology and coding-based modelling was conducted to explore the differences between more effective and less effective teams. Through its multiple methods EAST was used in order to analyse the teams' ability to communicate, cooperate, coordinate, and develop appropriate SA and schemata. Coding was then applied to the communication transcripts of the two teams to represent the interaction of these five factors during mission performance.

Inter-rater Reliability

As in Chapter 4, the results of the coding in this chapter were subjected to inter-rater reliability analysis in order to assess the level of homogeneity of the author's analysis and two independent raters' analysis. For the CDA, communication transcript coding and communication transcript link coding inter-rater reliability was assessed. Two independent raters were provided with a set of coding schemes and a 10 per cent sample of each piece of coding (the CDA and communication transcripts). A series of explanations were provided for each task step presented in the CDA and for each communication act within the communication transcripts. The explanation was provided in order to ensure that the independent raters could accurately comprehend the technical language included. The responses of the independent raters were recorded onto a response sheet and entered into a Microsoft Excel worksheet. A percentage agreement between the authors' initial coding and the independent raters' coding was calculated and is presented below in the relevant Findings section.

EAST Findings

Based upon the observation of the two teams' performance, a brief description is provided in order to give the reader a level of context around the analysis presented in this chapter. Two teams were chosen for comparison, based on their performance in the mission. Discussion with Subject Matter Experts, trainers at the institution, was used to identify a team that characterised more effective mission performance and a team that characterised less effective mission performance as described above under 'Methods'. The incident of fratricide, which occurred within the less effective team and did not occur within the more effective team, makes a clear distinction between the relative effectiveness of the two teams.

The more effective team began the mission by utilising the intelligence it had been given and assigning assets to search specific target areas, or to move ahead of the convoy, heading to the Shura to ensure that their route was secure. Assets provided regular reports, or SITREPs (SITuation REPorts), regarding the areas they had been assigned to, allowing the command team to develop a pattern of life, or an appropriate awareness of the situation as it developed. Each asset's assigned area was allocated a section on a board in the JFC which the command team used to annotate the SITREPs, providing a clear illustration of who saw what, where and when. In addition to this a target board was created in the JFC, built up from the assets' SITREPs. Each time a sighting was reported, it was placed on the board and assigned a specific code name; this code name was then communicated to all other assets, along with a grid reference to ensure a clear depiction of the target situation, and to ensure that no SITREPs were replicating information already passed. The command team also ensured that each sighting report was confirmed by a second asset. These boards, and confirmatory sighting reports, enabled the command team to identify and confirm targets effectively and to allocate resources appropriately, either to watch or to engage the targets. The strategy, watch or engage, and the assigned asset were annotated onto the board.

In this way, the more effective team was able to appropriately identify and maintain an accurate awareness of targets, and of their assets' locations and roles. When a request was passed from a neighbouring JFC to engage a fleeting target (a target moving rapidly into their area), the command team used the same methodical approach. They analysed their asset tracking board to identify an asset able to gain an appropriate view of the fleeting target. The asset was assigned, and confirmed that there was a group moving near the grid reported. As the grid was outside the JFC area of operations, they communicated efficiently with their headquarters (HQ) to request permission to engage the target. This communication resulted in the HQ checking the area and consequently passing back information that the fleeting target was in fact a friendly Special Task Force.

In the less effective team the distribution of information was less ordered, with multiple SITREPs being called in simultaneously. Confusion arose over target locations and descriptions, which led to the same target being given two conflicting code names, and separate targets being merged into one code name. The problems with consistent labelling and communication procedures meant that the command team did not have a clear awareness of the target situation, or of where their own assets were, or what they had been tasked with. In light of this, the command team was unable to make a decision to engage any targets with confidence and was under increasing pressure to take such decisions due to the mounting number of targets. At this point in the mission, the team received a request from a neighbouring JFC to engage a fleeting target (a target moving rapidly into their area). The command team deemed this to be reliable and accurate information and made the decision to engage the target. Due to the problems with awareness of their assets' tasking and locations, the team made the decision to use an asset which was unable to get a positive identification of the target. Without attempting to get a positive identification from another asset (which was available), or permission to engage outside their area, the command team engaged the fleeting target. This target was in fact a friendly Special Task Force.

Coordination and Cooperation

Hierarchical Task Analysis (HTA)   HTA involves decomposing a scenario into goals, sub-goals and tasks in order to understand the manner in which teams undertake the overall goal (Annett 2005). In the case of Operation Shura, an HTA was created for both the more effective and the less effective teams in the same manner as discussed in Chapter 4. From the HTA a summarised task network can be derived, an overview of the first two levels of the task networks for both teams can be seen below in Figure 5.2.

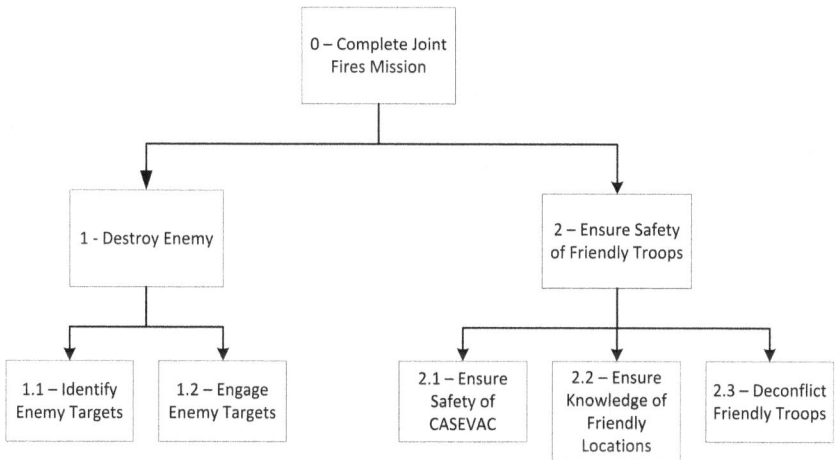

Figure 5.2   Overview of HTA for the more effective and the less effective teams

At this level the task networks are identical; however, closer examination of the full HTAs reveals that the two teams conducted the mission scenarios in quite different ways, despite both Batteries following the same SOPs, ROEs, etc.

Figure 5.3, on the following page, represents an overview of the sub-goals and tasks present in the HTA of both teams. There is a higher level of sub-goals and tasks within the less effective team when compared to the more effective team. As the level of overall goals is the same for the two teams (as presented in Figure 5.3) it can be argued that the more effective team was able to meet its goals in a more efficient and succinct manner. The notion that effective teams are able to manage work more effectively is supported within the wider literature (Orasanu 2005).

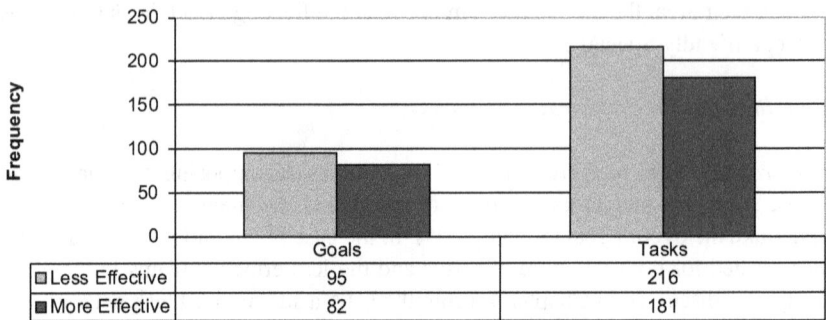

	Goals	Tasks
☐ Less Effective	95	216
■ More Effective	82	181

Figure 5.3   Comparison of sub-goals and tasks between the less effective and the more effective teams

In addition, when the sub-goals and tasks are examined in more detail a number of inappropriate tasks can be identified within the less effective team. An example of this is task 1.1.1.1.1.1.7; [Conduct Show of Force] in which a Show of Force was conducted to move away a Dicker (a single person watching for friendly forces then contacting enemy forces to arrange an attack). This Show of Force involved flying a fast jet directly over the ongoing Shura, even though all kinetic activity in the area had been specifically restricted by the higher levels of command. As was pointed out by the trainers at the institution, a fast jet flying over a Shura does not set the right ambience for the local population to trust the military or to maintain secrecy of the event from the 'enemy'.

Crucial steps were also missed out by the less effective team: for example task 1.2.1.4.1; [Maintain Eyes on] Fleeting target; and task 1.2.5.2.2; [Get Positive Identification], were not present. This means that a fleeting target was engaged across boundaries without any asset having eyes on the target during the engagement.

Coordination Demands Analysis (CDA)   CDA identifies each task step from the HTA as involving individual or teamwork; each teamwork task is then rated on a series of coordination dimensions to derive an overall coordination value for teamwork tasks (Burke 2005). The CDA undertaken in this chapter was subject to inter-rater reliability tests following the procedure outlined above under 'Methods'. The results of the reliability testing revealed a mean degree of agreement of 89 per cent in the analysis of the initial CDA. The results of the CDA analysis conducted for the more effective and the less effective teams are summarised in Table 5.1.

The CDA analysis revealed no considerable difference between the level of teamwork tasks undertaken by the more effective and less the effective teams. Although there is a slightly greater level of teamwork tasks within the less effective team, there is not a noteworthy greater percentage of teamwork tasks in relation to the overall tasks performed: 75 per cent in the more effective team and 74 per cent in the less effective team. In light of this the greater level of teamwork tasks

Table 5.1   Summary of CDA analysis

	Less effective	More effective
Total task steps	216	181
Total individual work	56 (26%)	46 (25%)
Total teamwork	160 (74%)	135 (75%)
Mean Total Co-ordination	2.4 (80%)	2.9 (97%)

within the more effective team may simply be representative of the greater level of tasks they conducted. However, when the teamwork tasks are examined in more detail, a difference between the coordination present within the teamwork tasks is identified.

Within the more effective team there is a higher level of coordination in the teamwork tasks: 2.9 (97 per cent) compared to 2.4 (80 per cent) in the less effective team. From this it can be argued that although there were not quantitatively more acts of coordinated behaviour in the more effective team, the coordinated acts they did perform involved a higher level of coordination. In conclusion, there was not a greater quantity of teamwork but there was a higher quality of teamwork in the more effective team. Coordination has been hypothesised to be positively correlated to effective team performance by researchers such as Stanton and Ashleigh (2000), Entin and Serfaty (1999) and Wilson, Salas, Priest and Andrews (2007) and its importance has also been emphasised within distributed environments as explored within this chapter (Fiore et al. 2003, Boiney 2007, Paris et al. 2000).

Communication and Cooperation

Communication Usage Diagram (CUD)   CUD identifies the frequency of communication acts that occurred in each team and the medium through which these acts were passed (Watts and Monk 2000). At a basic level, the number of communication acts that occurred differed widely between the more effective and the less effective teams, as is illustrated in Figure 5.4.

Again, as in Chapter 4, due to the use of only one communication medium, radio networks, a full CUD analysis is not undertaken within this chapter.

As is discussed in Chapter 4, research has frequently shown that a higher level of communication acts is positively correlated to effective performance (Svensson and Andersson 2006), but it is important to note that raw data does not always equate to useful information. Walker, Stanton, Salmon and Jenkins (2008) showed that a higher level of data transfer did not necessarily mean a higher level of task relevant, useable, information. From the results presented within this chapter it is proposed that although there were greater levels of communication acts within the less effective team, there was not a greater amount of useful information. In the same vein, the fact that the more effective team members were able to achieve

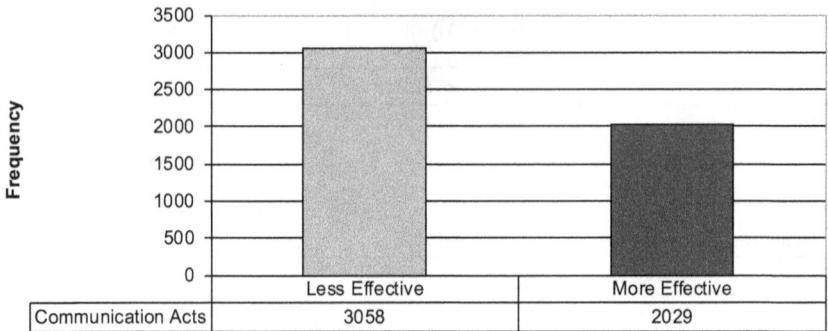

Figure 5.4   CUD summary

their mission goals using a lower level of tasks implies that they were able to pass the appropriate amount of information using fewer communication acts. As is the case here, past researchers have shown that a higher level of communication is not always positively correlated with more effective team performance (Orasanu 1995, Salas, Rosen et al. 2007). Moffat argues that there 'appears to be a point where the knowledge available to the Commander exceeds his capacity to act on it' (2003a: 130). In such situations the decision-maker must be able to identify and focus on the pertinent information, but this is not always possible. Boiney discusses a number of ways in which information may be present but not attended to, including:

> excessive distractions; getting lost in an overwhelming deluge of data, insufficient or excessive trust in particular information sources, confusion over priorities, failures of communication or memory, or critical interruptions. (Boiney 2007: 5)

Radio networks   To gain further insight into the communication structures employed by the more effective and less effective teams, the radio networks utilised by the two teams were explored. These radio networks and the frequency of their use, in both teams, are summarised in Figure 5.5.

Within the more effective team only three radio networks were used: one to talk to air assets; one for fire assets; and another to talk to the Company level. The less effective team used seven different radio networks (with a small number of communications being made on a further two networks). A radio network was used for Fires, and a further radio network was used specifically to let the FSTs talk to Mortars. Another was used for Air and then a further two radio networks were used by the FSTs to control the jets they had been assigned, and an additional radio network was used for Air to Air communications. The final radio network was used to talk to the Companies involved in the mission.

	Air 1	Air 2	Air 3	Air 4	Air 5	Company	Fires 1	Mortar	Fires 2
Less Effective	912	283	83	472	3	492	581	227	5
More Effective	932	0	0	0	0	443	654	0	0

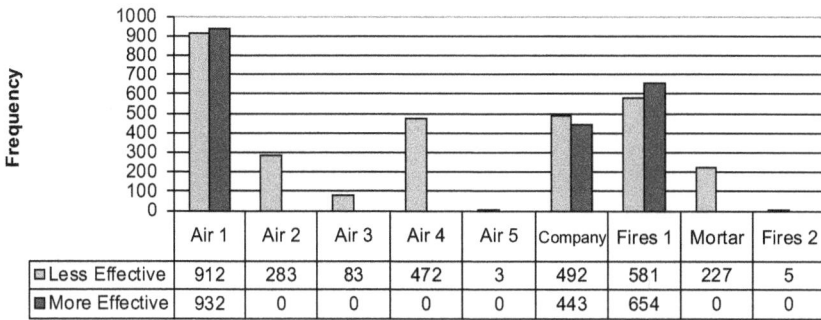

Figure 5.5   Communication usage over radio networks

From the analysis of radio networks it could be hypothesised that the less effective team had a greater potential to communicate with others, due to the greater number of radio network connections they utilised. At the start of the mission both teams had access to the same number of radio networks, but the less effective team chose to utilise a greater number of these radio networks than the more effective team. Although this initially sounds positive, research suggests that an 'open' communication system is not always correlated with effective performance. Research by Omodei, Wearing and McLennan (2000) explored the performance of Command and Control teams undertaking a 'fire chief' scenario within an 'open' communication structure and a 'restricted' communication structure. The results of their study revealed that an open communication structure was correlated with a greater level of importance placed on information processing; they posited that this was indicative of an increase in information processing demand within the open system. Omodei, Wearing and McLennan (2000) cite research by Brehmer and Svemark (1994) and Wang et al. (1991), in which 'restricted' communication structures were correlated with more effective performance than 'open' communication structures. This research supports those findings and further analysis emphasises the point, as illustrated in the Social Network Analysis below.

Social Network Analysis (SNA)   Further exploration of communication was conducted through SNA. SNA identifies every communication act that occurs and creates a matrix of from–to communications. From this analysis a number of statistics can be identified that reveal details of the communication structure of the two systems. Five such metrics are discussed here: density; cohesion; diameter; centrality; and sociometric status:

1. Density: represents the level of interconnections within a network (Walker, Stanton, Kazi et al. 2009, Walker, Stanton, Salmon, Jenkins et al. 2010).

2. Cohesion: represents 'the number of mutual connections in the network divided by the maximum possible number of such connections' (Benta 2003).
3. Diameter: represents 'the largest number of [elements] which must be traversed in order to travel from one [element]' (Weisstein 2008, Harary 1994).
4. Centrality: represents how closely an individual is connected to other individuals in the network: the distance between them (Houghton, Baber, McMaster et al. 2006).
5. Sociometric Status: represents how busy an individual is; how much an individual communicates (Houghton, Baber, McMaster et al. 2006).

Density and cohesion
Figure 5.6 below represents the density and cohesion figures for both the more effective and less effective teams.

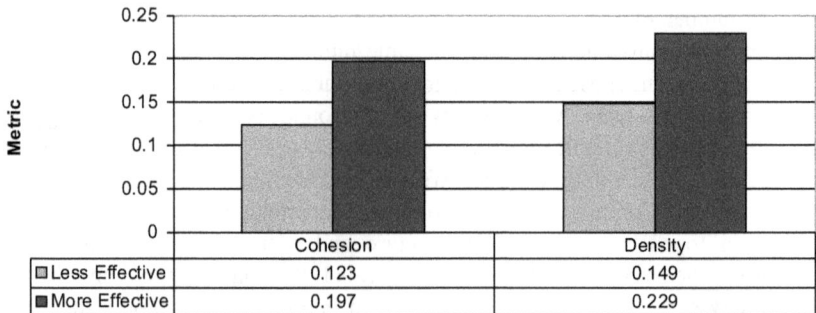

	Cohesion	Density
☐ Less Effective	0.123	0.149
■ More Effective	0.197	0.229

Figure 5.6   Density and cohesion values

The analysis reveals a higher level of density and cohesion in the more effective team compared with the less effective team. A higher level of cohesion represents a higher level of direct, informal links between individuals in the team. Higher levels of density represent a greater number of links between individuals relative to the number of possible links in the system (Walker, Stanton, Salmon, Jenkins et al. 2010, Houghton, Baber, McMaster et al. 2006). These metrics portray the more effective system as having a greater level of relationships between individuals in the system compared to the less effective system.

Diameter
Both teams had the same level of diameter: 4.

Sociometric Status
Table 5.2 summarises the results of the analysis of sociometric status within the two teams.

Table 5.2   Summary of sociometric status values for the more effective and the less effective teams

	Less effective	More effective
Min	0.04	0.13
Max	63.72	47.77
sum	244.64	184.45
Mean	9.40	8.11
Variance	226.45	118.26
Standard deviation	15.04	10.87

Comparison of the sociometric status values for the two teams highlights a number of differences. The less effective team has a greater range of values – with a lower minimum value and a higher maximum value and, subsequently, a larger variance and larger standard deviation compared with the more effective team. This suggests that there is a wide variance between the individuals in the level of contribution they are making to the communication flow of the network, with some individuals contributing a lot and others barely contributing at all, whereas in the more effective team the variance is lower, suggesting that the contribution levels had a greater degree of homogeneity across the individuals.

From the analysis it can be posited that the greater frequency of communication, on more communication channels could have contributed to the incident of fratricide.

Centrality
Table 5.3 below presents a summary of the analysis of centrality in the more effective and the less effective teams.

Table 5.3   Summary of centrality analysis

	Less effective	More effective
Min	8.76	7.30
Max	1087	833
Sum	1422.41	1087.61
Mean	54.70	47.28
Variance	42639.8	28069.1
Standard Deviation	206.49	167.53

This summary table illustrates a greater level of centrality in the less effective team, with a higher mean and a higher sum value compared with the more effective team. This means that the individuals in the less effective team were more connected. This coincides with the earlier CUD analysis, which highlighted a greater number of radio networks – communication channels – within the less effective team. The less effective team also had a greater range of centrality across its individuals with a much higher maximum value and a higher variance and standard deviation than the more effective team.

Summary of Social Network Analysis
The SNA has revealed higher levels of sociometric status and centrality in the less effective team. In contrast to this, the more effective team had higher levels of cohesion and density. This suggests that the higher levels of sociometric status and centrality in the less effective team represent unnecessary communications that did not need to occur and served only to overcomplicate, overload and possibly confuse the individuals in the team. In the more effective team the communication acts undertaken and the communication structure were both leaner, but could be said to be more efficient and more effective, leading to higher levels of cohesion and density.

Overall, this shows that there was a higher level of communications and a wider communication network in the less effective team. From the SNA analysis it could be surmised that a lower level of communication and a smaller communication network may be correlated with more effective performance. The notion that lower levels of communication may be beneficial is supported by research such as Urban, Bowers et al. (1995), which identified that effective teams were more able to efficiently communicate only appropriate information. Stanton, Stewart et al. (2006), Dismukes, Loukopoulos and Jobe (2001) and Hutchins, Hocevar and Temple (1999) all highlight the detrimental aspects of excessive communication, suggesting that additional communications can be irrelevant and cause unnecessary cognitive workload.

The results of the Social Network Analysis support the assertion made earlier that excessive communication can mask relevant and appropriate information, making it harder for individuals to decipher and act on the most important communications.

Communication

In order to explore the communications within the mission scenarios further, the transcripts from both teams were coded using a grounded theory approach (Glaser and Strauss 1967). As in the analysis presented in Chapter 4, the coding scheme was drawn out from the data following Glaser and Strauss (1967) Open Coding framework

An example of an extract from the communication transcripts is presented here:

Hello Guns4 this is FST2. Send locstat over.

This extract was taken from the less effective team and illustrates the Fire Support Team requesting the location of the artillery unit 'Guns4'. This extract was coded as *request information*.

In order to illustrate the reliability of the communication transcript coding, inter-rater reliability tests were performed following the procedure outlined above under 'Methods'. An agreement level of 80 per cent was established between the three versions. Table 5.4 below represents a summary of the coding for the more effective and less effective teams, with both total numbers and percentage of the total communication in each team, presented.

Table 5.4 highlights four prominent differences in the communication types between the teams: *radio checks*; *requests to speak*; *send information*; and *send orders*. These differences are calculated based on the difference between numerical and percentage figures as a fraction of the overall communication acts within each team. The more effective team had a higher level of *radio checks*, despite the lower level of radio networks utilised by the team compared to the less effective team. This suggests that the individuals within the more effective team were more aware of the need to conduct regular radio checks.

Table 5.4   Results of coding for the more effective and the less effective teams

	Less effective n (%)	More effective n (%)
Acknowledge	607 (19.95%)	425 (20.7%)
Confirm	24 (0.79%)	10 (0.48%)
Radio check	25 (0.82%)	98 (4.77%)
Read-back	184 (6.05%)	140 (6.82%
Request	8 (0.26%)	4 (0.195%)
Request acknowledgement	112 (3.68%)	65 (3.17%)
Request confirm	1 (0.03%)	1 (0.05%)
Request granted	122 (4.01%)	51 (2.48%)
Request information	341 (11.28%)	272 (13.25%)
Request repeat	88 (2.89%)	29 (1.41%)
Request to speak	189 (6.21%)	71 (3.46%)
Request task	4 (0.13%)	0 (0%)
Send information	1189 (39.09%)	709 (34.53%)
Send order	51 (1.68%)	118 (5.75%)
Wait out	84 (2.76%)	50 (2.44%)
Wrong	11 (0.36%)	10 (0.48%)

The higher level of *send information* codes present within the less effective team: 39 per cent of overall communications compared to 34.5 per cent of the overall communications in the more effective team, presents an interesting finding. As is discussed earlier in this chapter, a greater level of information transferred does not always equate to a greater awareness, especially if the information sent is inaccurate. This finding is discussed in more detail below under 'Situation Awareness'.

There was a greater percentage of *requests to speak* codes in the less effective team in relation to the overall communication acts within that team. This may suggest that they had a more formalised radio procedure in which they asked permission before they spoke. The increased level of *requests to speak* may also have been due to the greater radio traffic and greater number of radio networks within the less effective team, with the increased level representing the individuals trying to ensure other individuals were free and listening before they relayed a message.

The greater level of orders sent in the more effective team: 5.75 per cent (of the total communication acts within the more effective team) compared with 1.68 per cent (of the total communication acts within the less effective team) in the less effective team, suggests that a greater level of guidance was provided by higher levels of command within the more effective team. Previous research has supported the assertion that communication of command intent is vital to success in complex situations (Cahillane et al. 2009), for effective team performance, and superior quality decision-making (Bryant 2006). Researchers such as Shattuck and Woods argue that subordinates 'must understand the supervisors' underlying intent with regard to plans and procedures', especially when supervisors and subordinates are distributed (2000: 280). Pigeau and McCann also emphasise this, stating that communication of intent is a:

> Commander's primary mechanism for initiating and maintaining goal directed action among subordinates. (2000: 166)

Situation Awareness

In Chapter 4, the utility of Leximancer to explore both SA and schemata at a systems level was illustrated. Within this chapter, the Leximancer software was also used to provide insights into the differences between the less effective and the more effective teams.

Information networks   All communication acts that occurred during mission completion were transcribed and inputted into the Leximancer software. Emergent concept seeds – information elements – were investigated for any inapplicable or repeated concepts. The concept seeds were then reduced to represent the core concepts for each network and linked based on their relational and conceptual links. The graphical output of the Leximancer analysis for both the more effective and the less effective teams is presented in Figure 5.7 and Figure 5.8.

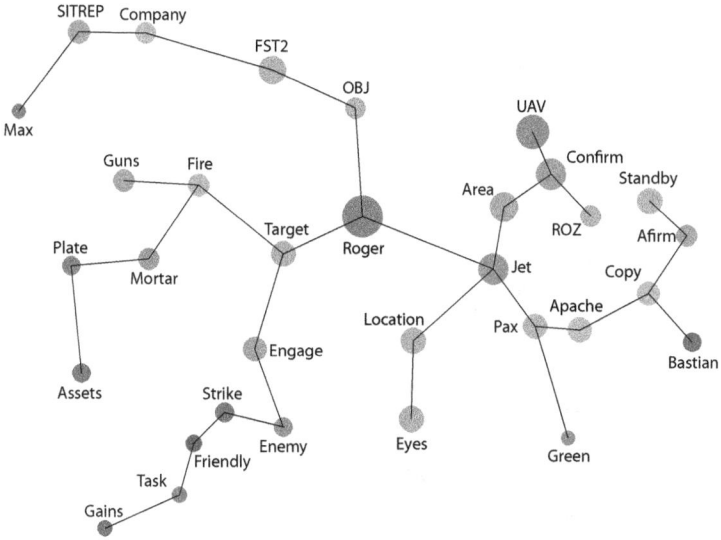

Figure 5.7   Less effective team information network

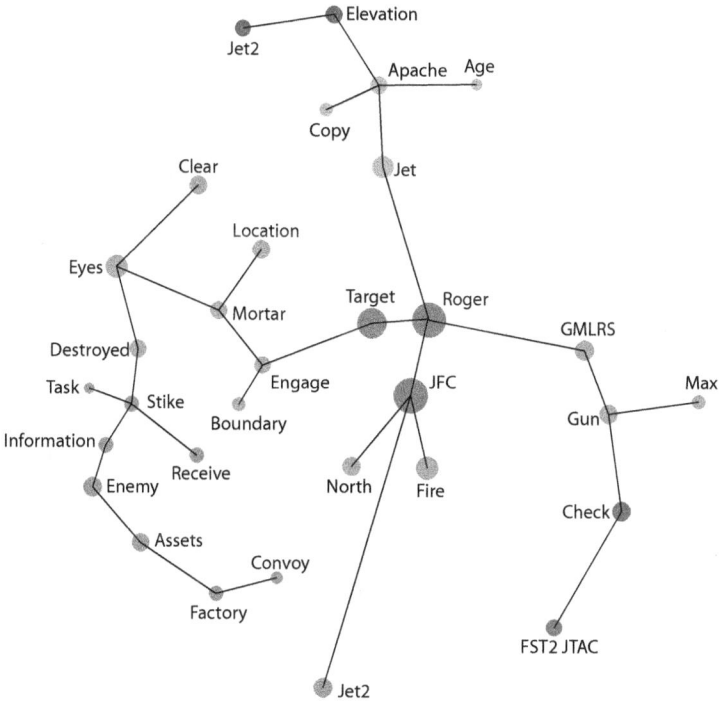

Figure 5.8   More effective team information network

Initial examination of the networks appears to illustrate a greater level of relationships between elements in the more effective team but to show that these relationships are less well defined, as is illustrated by the larger distance between elements compared with the less effective team.

Table 5.5 presents the core information elements and their relevance within the less effective and the more effective teams. The shaded cells represent elements present within both the more effective and the less effective teams.

Although there is a degree of overlap in the elements present in the teams, with 56 per cent of the elements present in both scenarios, there are also several differences: 44 per cent of the elements are present in only one of the teams. In addition to this, there are differences in the importance attached to and the relevance, or connectivity, associated with the elements between the two teams. From examination of Table 5.5, differences between the connectivity of elements is identifiable but the specific nature of these differences is unclear. In order to explore the connection of information elements within the network as a whole, graph theory metrics were derived for the networks developed through the Leximancer software, including: sociometric status; centrality; density; cohesion; and diameter. The results of this analysis are presented in Table 5.6.

The results of this analysis illustrate a higher level of centrality in the more effective information network when compared with the less effective information network.

The value of the diameter metric is lower within the less effective team, suggesting that the information present within the less effective team network is less distributed, with shorter paths between elements compared with the more effective team. This contrast can be explained by the DSA perspective and the wider Network Enabled Capability (NEC) literature, both of which argue that not everyone needs to know everything (Stanton, Stewart et al. 2006, Stanton, Salmon, Walker and Jenkins 2009a, 2009b, Salmon, Stanton, Walker and Jenkins 2009, Alberts and Hayes 2003).

> Each individual in each situation has a different need to information and can tolerate a different degree of ambiguity. (Alberts and Hayes 2003: 77)

The greater level of role and task diversity within the large JFC team of teams may mean that across the system as a whole different information is being used by different people in different ways and, therefore, paths between information elements vary depending on who is accessing the information. This leads to a more distributed set of information elements in the more effective team compared to the less effective team when people are not collocated (contrary to the finding for collocated teams in Chapter 4). It is suggested that in the more effective team, agents are aware that they do not require all information and therefore do not require a tightly integrated network of information.

According to Stanton, Stewart et al. (2006) DSA is focused on interactions between agents and they argue that each agent holds unique SA developed through

Table 5.5   Elements and relevance for the more effective and the less effective teams

Less effective			More effective		
Element	Count	Relevance	Element	Count	Relevance
roger	241	100%	JFC	173	100%
jet	149	62%	roger	173	100%
confirm	132	55%	target	134	77%
UAV	131	54%	jet	80	46%
FST2	118	49%	GMLRS	79	46%
area	112	46%	eyes	76	44%
fire	103	43%	fire	70	40%
pax	100	41%	gun	63	36%
apache	98	41%	company	58	34%
copy	95	39%	apache	53	31%
target	91	38%	location	51	29%
standby	90	37%	check	48	28%
eyes	87	36%	enemy	47	27%
engage	86	36%	assets	44	25%
location	83	34%	engage	42	24%
SITREP	79	33%	north	42	24%
objective	77	32%	clear	35	20%
company	77	32%	FST2	33	19%
guns	61	25%	copy	33	19%
affirm	58	24%	mortar	28	16%
mortar	58	24%	jet2	23	13%
ROZ	55	23%	strike	21	12%
plate	52	22%	elevation	19	11%
enemy	48	20%	destroyed	18	10%
assets	46	19%	convoy	18	10%
strike	26	11%	information	17	10%
Bastian	26	11%	max	17	10%
friendly	24	10%	receive	15	9%
gains	19	8%	boundary	14	8%
green	19	8%	factory	13	8%
max	19	8%	age	8	5%
task	13	5%	task	8	5%

Table 5.6   Information network metrics

	Less effective	More effective
Density	0.031	0.031
Mean sociometric status	0.625	0.625
Mean centrality	27.7	31
Diameter	12	15

transactions with other agents. SA is distributed across a system, so that one would expect the elements of that SA to be distributed. The way in which information is activated does not follow a 'shared' pattern and therefore paths are not tightly defined. Stanton, Stewart et al. (2006) argue that individuals have different views on systemic SA; they activate different elements at different times in different ways, creating a distributed network of information. The research presented within this chapter suggests that the diameter metric supports this proposition.

Exploring fratricide from a systems perspective forces researchers to explore not only multiple factors but also the way in which they interact, both quantitatively and qualitatively. In order to explore this further, an in-depth examination of the connections between elements in the two teams was undertaken. The results of this thematic analysis are presented below in Figure 5.9 and Figure 5.10 opposite.

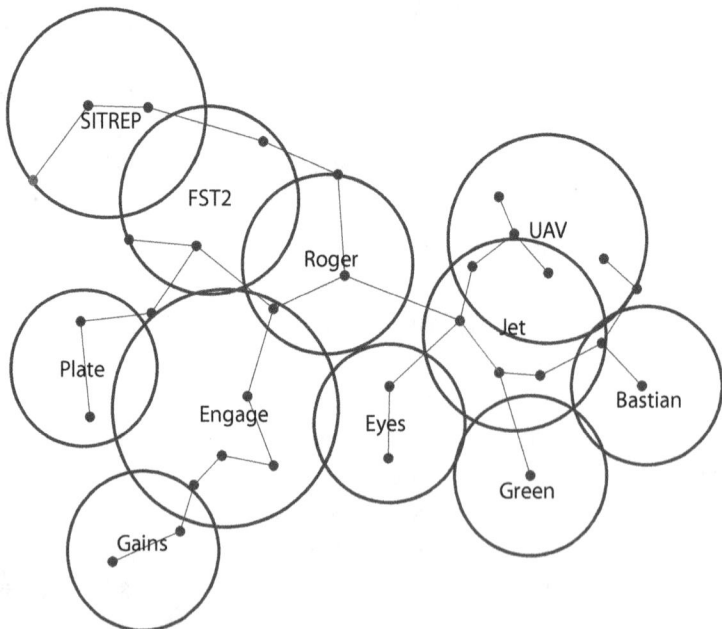

Figure 5.9   Less effective team themes

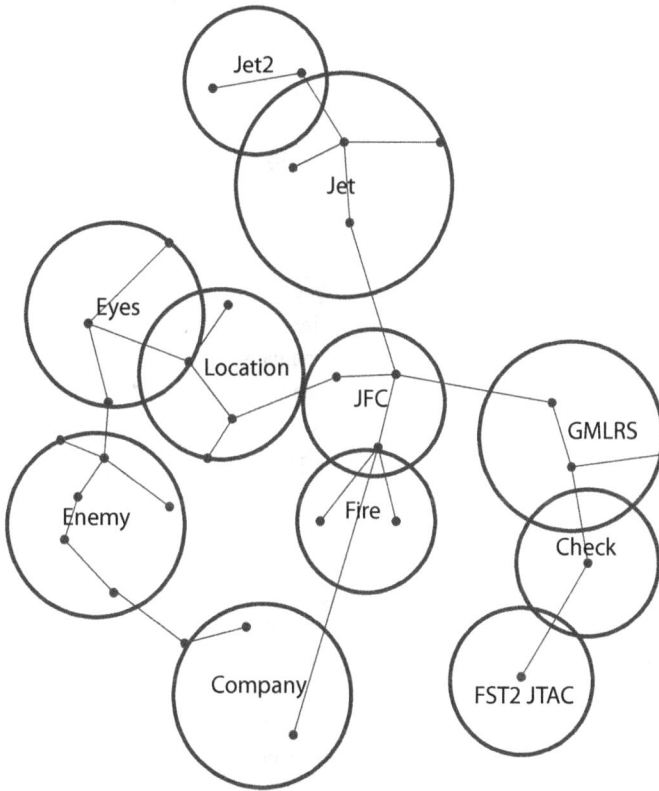

Figure 5.10   More effective team themes

The graphical illustrations of the thematic analysis above depict a number of differences between the analyses of the two teams. There are a number of different themes present in the two teams and the themes that do occur in both teams are connected in different ways. Table 5.7 represents the themes and the connectivity of the themes for the less effective and the more effective teams. The shaded cells represent themes that are present in both teams.

In Chapter 4, it was shown that information appeared to be incorrectly interpreted to a higher degree by the less effective team compared with the more effective team. In-depth analysis of the transcripts illustrates that the same is true in this case study. A description of the most prominent schemata and their impact on both teams is presented below, as derived from the themes presented in Table 5.7 and their relationships with the communication transcript.

In the more effective team the key themes of *check* and *location* illustrate a greater emphasis on ensuring accurate location information. These themes were not present in the less effective team. *Eyes* also had a slightly higher level of connectivity within the more effective team compared with the less effective team

Table 5.7   Themes for more effective and less effective teams

Less effective		More effective	
Theme	Connectivity	Theme	Connectivity
jet	1	JFC	1
UAV	0.86	enemy	0.41
roger	0.65	jet	0.39
FST2	0.45	location	0.34
engage	0.42	GMLRS	0.33
eyes	0.3	eyes	0.32
SITREP	0.24	fire	0.25
plate	0.16	company	0.18
gains	0.06	check	0.1
Bastian	0.05	jet2	0.1
green	0.03	FST2	0.07

(32 per cent compared to 30 per cent), this refers to an asset being 'eyes on', which means that they are in direct observation of a target or area. From this it could be suggested that the more effective team took a more vigilant approach to the mission, as is illustrated by the following extract from a communication transcript:

> there are probably enemy in the area so we need to keep our eyes open and focus on possible leads to identify all enemy forces present.

In the less effective team the theme *engage* was prominent (42 per cent connectivity) but was not present in the more effective team. The similar theme of *fire* was present within the more effective team but had a lower level of connectivity (25 per cent). From this it could be suggested that the less effective team had a greater focus on engaging than the more effective team.

Within the less effective team the theme with the highest level of connectivity was *jet*, which had 100 per cent connectivity, compared to only 39 per cent connectivity in the more effective team. In the more effective team the theme *JFC* had the highest level of connectivity (100 per cent), but was not present within the less effective team. This difference is illustrative of different approaches to the scenario by the more effective and the less effective teams. Members of the more effective team were focused on what one another were doing, on coordinating within the *JFC*, whereas members of the less effective team were focused on strike assets – such as the *Jet* – representing a focus on engagements over coordination.

The results of the thematic analysis suggest that the more effective team approached the scenario in a more cautious manner, triggering schemata regarding accurate identification and detailed information transfer. In comparison, the less effective team appear to have approached the scenario focusing upon engagements, placing less emphasis on sightings and accurate location reports.

The analysis of the information elements in the less effective and more effective teams illustrated that, possibly as a by-product of the excessive levels of information transferred within the less effective team, information integration was inappropriate. A higher level of integration was found within the less effective team, but examination of this connectivity revealed multiple instances of inappropriate information integration. From this it could be suggested that within the less effective team the greater level of information passed meant that team members were unable to integrate this information appropriately into their appreciation of the situation, leading to inappropriate schemata being triggered.

Summary of results   A recurring theme in the EAST analysis of this case study is the more efficient succinct manner in which the more effective team carried out the mission, with respect to task steps, communication and information. It appears that the less effective team over complicated the mission, taking a greater amount of tasks to reach the same goal. They also appear to have over-communicated, sending irrelevant or unnecessary information. The results of the EAST analysis are summarised below in Table 5.8.

Table 5.8   Summary of findings

		Less effective	More effective	
	HTA	Higher number of tasks	Lower number of tasks	Coop and Coord
	HTA	Higher level of inappropriate tasks	Lower level of inappropriate tasks	
	HTA	High number of crucial tasks omitted	No or low number of crucial tasks omitted	
	CDA	Similar level of teamwork tasks	Similar level of teamwork tasks	Coord
	CDA	80% – Lower level of coordination	97% – Higher level of coordination	
EAST	CUD	3058 – Higher rate of communication acts	2029 – Lower rate of communication acts	Comms
	CUD	9 – Higher rate of communication networks/channels	3 – Lower rate of communication networks/channels	F3 Model
	SNA	Lower levels of density and cohesion	Higher levels of density and cohesion	Comms and Coop
	SNA	Higher levels of sociometric status and centrality	Lower levels of sociometric status and centrality	
	IN	Higher level of incorrect interpretation of information	Lower level of incorrect interpretation of information	SA and Schemata
	IN	Lower level of centrality	Higher level of centrality	
	IN	Higher level of information connectivity	Lower level of information connectivity	

Interaction of Factors

In order to explore the manner in which the factors interact with one another the F3 model was populated for both teams. The population of the links between the models was created through coding of the transcripts. The coding scheme was developed within Chapter 1 using a grounded theory approach and was presented in Chapter 2. Each transcript was coded for both more effective and less effective links between the F3 model's factors. Two examples of how these links are identified can be seen in the following two extracts:

> JFC, FST1: Convoy update – we are now passing grid *** still moving north east over.

The first extract, above, is taken from the more effective crew's transcript and has been labelled as representing the link *Coordination – Situation Awareness*. The Fire Support Team's coordination with the Joint Fires Cell through their update enabled the Joint Fires Cell to have a correct SA regarding their assets' movements:

> FST1 this is Company 1: Is that your call sign now firing to my north.

The second extract, above, is taken from the less effective crew's transcript and has been labelled as representing a breakdown in the link *Coordination – Situation Awareness*. Due to a lack of coordination between the friendly call signs involved in the mission, the less effective crew have a poor SA regarding engagements occurring very close to their location.

Coding of the communication transcripts in this way enabled a representation of the positive links between factors within the model, as well as identifying breakdowns in the links between the factors. In order to ensure the reliability of this coding, inter-rater testing was undertaken following the procedure outlined above under 'Methods'. A percentage agreement of over 80 per cent was found.

The populated F3 models for the less effective and the more effective crews, comparing both positive links between factors as well as negative links, are presented in Figure 5.11.

Comparing the two teams' positive links side by side reveals a number of core positive links present in both teams. Three links have a dramatically high level of occurrence in both teams: Communication to Situation Awareness; Communication to Coordination; and Coordination to Situation Awareness. From this it can be concluded that key aspects of more effective decision-making are present within both teams, but at a slightly higher level in the more effective team.

Examination of the negative links in the less effective team reveals that the most important (highest frequency) negative link in this model is the link between Communication and Coordination. Although the less effective team had a high level of positive links between these two factors, the model above portrays a high level of negative links as well.

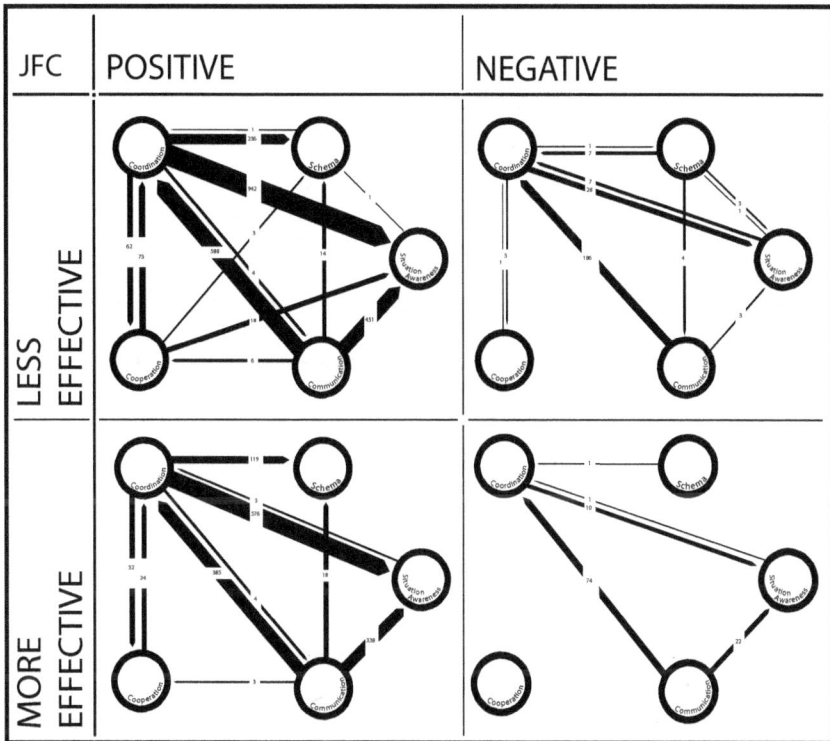

Figure 5.11   **Positive and negative links for the more effective and the less effective teams**

In summary, the link between Communication and Coordination has the highest frequency for negative links, and the links between Communication, Coordination and Situation Awareness appear to have the highest frequency for positive links. From this it can be posited that the links between Situation Awareness, Coordination and Communication represent key relationships in incidents of fratricide.

An additional finding from the comparison of the populated models is that the presence of positive links seems to have less of an effect than the presence of negative links. This means that more positive links cannot compensate or counteract negative links. Table 5.9 and Table 5.10 show the frequency and presence of links in each of the four models.

With respect both to frequency (the total number of links) and type (the number of different links between factors) it appears that the presence of 'negative' links has a bigger effect than the presence of 'positive' links. The less effective team have a greater frequency and type of positive links, but also a greater frequency and type of negative links, indicating the disproportionate impact of negative links on the team's performance.

Table 5.9   The frequency of positive and negative links in both teams

Frequency	Positive	Negative	Difference
Less effective	2399	243	2156
More effective	1521	108	1413
Total	3920	351	

Table 5.10   The presence of positive and negative links in both teams

Presence	Positive	Negative	Difference
Less effective	13	11	2
More effective	10	5	-5
Total	23	16	

Summary

There are four core findings derived from the research presented in this chapter:

1. Efficient communication strategies are correlated with effective team performance.
2. Efficient task-work is correlated with effective team performance.
3. Efficient information transfer is correlated with effective team performance.
4. Coordination is a necessary mediatory factor between Communication and Situation Awareness.

These results are supported by wider literature, as discussed in Chapter 1. The finite limits of team members and teams to process information (Reason 1990, Woods et al. 1994, Maule 2009) means that teams are unable to process large amounts of information and, consequently, important information can become masked (Stanton, Stewart et al. 2006). This limit is further emphasised within distributed teams where additional cognitive resources are required to coordinate across the team (Fiore et al. 2003), emphasising the need for efficient communication strategies (Urban et al. 1995, Gorman, Cooke and Winner 2006, Dismukes, Loukopoulos and Jobe 2001, Hutchins, Hocevar and Kemple 1999). Additionally, in military situations bandwidth limitations further necessitate the need to ensure efficient communication and information transfer (Walker, Stanton, Jenkins et al. 2009, 2010). More efficient communication strategies allow the transfer of relevant information, without overloading the individual with inappropriate information or masking the relevant information they need to process in order to achieve effective performance.

Conclusion

More effective and less effective military mission performance from a Joint Fires training facility were compared using the EAST methodology to draw out key features of the decision-making process and further the development of a model of fratricide, the F3 model. The observation of an Army and RAF Joint Fires mission enabled additional development of the F3 model and the illumination of a series of interesting findings. Through the application of the EAST methodology to both a less effective and a more effective completion of this Joint Fires mission, the F3 model's core factors were further validated as key in the incidence of fratricide.

The analysis presented within this chapter revealed that fewer communications between teams leads to more effective performance. This may be due to the distributed environment and the number of people communicating with one another. In this situation, there is only a certain amount of communications any team can process, so it is important to be concise. Secondly, due to the distribution of tasks, not everyone needs to know everything; communications should, therefore, be team-specific, supported by the research surrounding DSA (Stanton, Stewart et al. 2006, Salmon, Stanton, Walker and Jenkins 2009). Additionally, the importance of coordination has been identified within this multi-team environment in line with research surrounding team decision-making (Fiore, Salas et al. 2003). The ability to map the results onto the research surrounding DSA and coordination literature provides validation for the analysis and the utility of the EAST method to explore team decision-making.

In conclusion, the results illustrate the impact of communication-masking in less effective teams. Masking is present in the coding of communication type, with the negative links masking the effects of the positive links; there is further masking in the communication frequency analysis, with a greater level of communications masking the important/relevant communications; and, finally, there is masking in the information networks, with inaccurate or incorrectly interpreted information masking the accurate information. At all levels of the system this multi-faceted phenomenon can be seen. This is important to the military for numerous reasons, not least the bandwidth problem. Military personnel operate in environments with drastic consequences and need to ensure that the right information and the right communications go to the right people through the right communication pathways, and that valuable bandwidth, technological, cognitive and team resources are not wasted (Walker, Stanton, Jenkins et al. 2009, 2010).

Chapter 6
Is it Better to Be Connected?

Introduction

Chapter 5 explored fratricide within a Joint Fires Cell (JFC) training scenario. The analysis revealed that coding and EAST analysis were able to reveal differences between an effective mission completion and a mission completion involving an incident of fratricide. The results provided further support for the F3 model and its ability to explore the Human Factors issues associated with these incidents. Within this chapter a further case study of fratricide is explored, this time within a Fire Support Team (FST) scenario. The combination of EAST and coding-based modelling was again used in order to explore the similarities and differences between two teams' completion of a Close Air Support mission, one effective and one resulting in an incident of fratricide.

Distributed Situation Awareness (DSA)

The previous chapter emphasised the differences in communication between the less effective and the more effective teams and suggested that these differences could be explained from a DSA perspective, arguing that in large diverse teams not everyone needs to know everything. Building on this, this chapter provides a further exploration of the applicability of the theory of Distributed Situation Awareness. According to Stanton, Stewart et al. (2006) the theory of DSA is built on six key principles:

1. SA is held by human and non-human agents.
2. Agents within a system hold different perspectives of the same scene.
3. The compatibility of SA between agents is dependent upon their tasks/ roles.
4. Agents within a system use both verbal and non-verbal communication.
5. SA holds loosely coupled systems together.
6. Agents are able to compensate for degraded SA in other agents.

The main principle of DSA is that people interact with the world around them and DSA is a resultant emergent property of this interaction (Stanton, Stewart et al. 2006, Stanton, Salmon, Walker and Jenkins 2009a, 2009b, Salmon, Stanton, Walker and Jenkins 2009). Due to its inception in interactions, DSA does not exist within an individual, but rather is distributed across a system (Stanton, Stewart et al. 2006, Stanton, Salmon, Walker and Jenkins 2009a, 2009b, Salmon Stanton,

Walker and Jenkins 2009). DSA is a systemic property, so that individuals within the system are able to extract information from this system-wide SA via exchanges with other agents in the system. The way in which this information is then interpreted by the individual depends upon the schemata. Schemata are developed from individual experiences and roles and are therefore distinct to each individual (Stanton, Stewart et al. 2006, Stanton, Salmon, Walker and Jenkins 2009a, 2009b, Salmon, Stanton, Walker and Jenkins 2009, Neisser 1976). Advocates of DSA strongly oppose the notion of shared SA, as is discussed in Chapter 1, for two core reasons. Firstly, individuals hold unique schemata based upon their own experience with the world, as such information, even the same piece of information, is interpreted by different individuals in different ways (Stanton, Stewart et al. 2006, Stanton, Salmon, Walker and Jenkins 2009a, Salmon Stanton, Walker and Jenkins 2009). Secondly, members of a team, or system, do not have the same roles and therefore do not require the same information, or do not require the same piece of information for identical purposes (Stanton, Stewart et al. 2006, Stanton, Salmon, Walker and Jenkins 2009a, Salmon, Stanton, Walker and Jenkins 2009).

Stanton, Salmon, Walker and Jenkins (2009a) argue that SA consists of two core elements: compatible SA and transactive SA. Individuals within a system interpret information in a unique manner, due to their individual schemata; however, this SA can be both unique and *compatible* (Stanton, Stewart et al. 2006, Stanton, Salmon, Walker and Jenkins 2009a, 2009b, Salmon, Stanton, Walker and Jenkins 2009). In a successful team, team members will access information, interpret it individually, and the resulting perception will be compatible with that of other team members, in that 'it is collectively needed for the team to perform the collaborative task successfully' (Stanton, Salmon, Walker and Jenkins 2009a: 65). Transactive SA refers to the manner in which individuals within a system access appropriate SA. Salmon et al. (2009) found that members of a team do not share SA, instead they engage in transactions during which SA is exchanged between team members. This transaction is a cyclical process:

> The act of passing awareness onto another agent serves to modify the receiver's situation awareness. Both parties are using the information for their own ends, integrated into their own schemata and reaching an individual interpretation. Thus the transaction is an exchange rather than a sharing of awareness. (Salmon, Stanton, Walker and Jenkins 2009: 208)

Individuals engage in SA transactions and utilise this information in different ways, for different purposes, but in a compatible manner (Stanton, Stewart et al. 2006, Stanton, Salmon, Walker and Jenkins 2009a, 2009b, Salmon, Stanton, Walker and Jenkins 2009). The focus of DSA is how agents within a system successfully conduct transactions and develop compatible interpretations of information (Stanton, Stewart et al. 2006, Stanton, Salmon, Walker and Jenkins 2009a, 2009b, Salmon, Stanton, Walker and Jenkins 2009). Good SA is defined as 'activated knowledge for a specific task within a system at a specific time by specific agents' (Stanton, Salmon, Walker and Jenkins 2009b: 6).

The theory of DSA has been successfully applied to numerous domains, including aviation (Stewart et al. 2008), energy distribution (Salmon, Stanton, Walker and Jenkins 2009), Command and Control (Stanton, Jenkins et al. 2009) and anaesthesia (Fioratou et al. 2010). Within this chapter the applicability of, and implications for, the theory of DSA within incidents of fratricide is explored.

Method

Scenario

The case study analysed within this chapter explores two Fire Support Teams undertaking pre-deployment Close Air Support training. Fire Support Teams are defined as teams responsible for:

> directing artillery fire and close air support (ground attack by aircraft) onto enemy positions. Because artillery is an indirect-fire weapon system, the guns are rarely in line-of-sight of their target, often located tens of miles away. The observer serves as the eyes of the artillery battery, calling in target locations and adjustment. (ARRSE 2010)

Due to the integration of air and ground assets, the military has developed specialised facilities in which the Army and the RAF can train together before they are deployed. The mission described in this chapter took place at a purpose-built training facility containing high-fidelity aircraft simulators and environments designed to simulate Fire Support Team positions and Joint Fires Cell (JFC) positions.

The Fire Support Team case study took place within a larger system which consisted of a Joint Fires Cell, two Fire Support Teams, a ground Company, an Unmanned Aerial Vehicle (UAV), an Apache helicopter, two fast jets, a GMLRS platform (Guided Multiple Launch Rocket System), a gun platform, a Chinook helicopter providing Casualty Evacuation (CASEVAC), a Mortar shell platform and Airspace Control Main (whose role it was to provide Airspace control for the wider area the FST were fighting in). The structure of the assets is represented in Figure 6.1.

This chapter presents a focus on the FST, as opposed to the JFC focus of Chapter 5. As illustrated in Figure 6.1, the FST team sits at a lower systemic level than the JFC, providing an exploration of an additional level within the military system. The mission analysed was named 'Just another day in Kajaki' and the FSTs were tasked with undertaking patrols in the area of Kajaki in order to maintain peace in the area. The pre-mission brief stated that there was no new intelligence regarding threats or enemy activity, reporting that it was just another day in Kajaki.

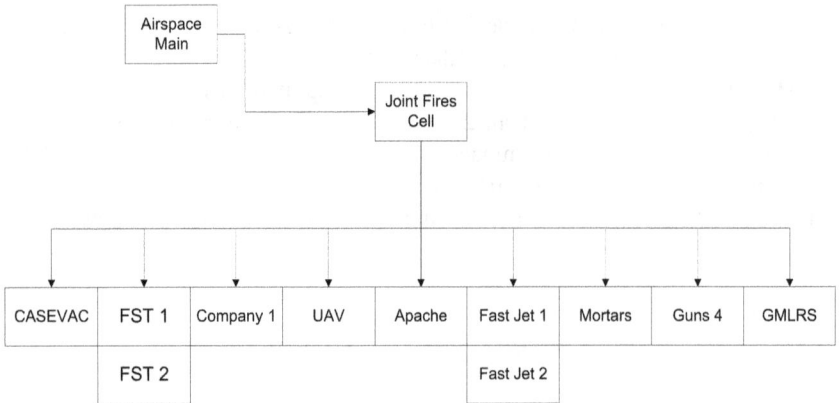

Figure 6.1   Asset structure within the mission

During the mission FST 1 was able to identify a number of enemy targets effectively and plan efficient and safe engagements of these targets; this team will be called the more effective team. FST2 did undertake a less successful completion of the mission scenario; they were unable to identify all enemy targets and mistakenly tasked an Apache helicopter air asset to engage their own location while intending to engage an enemy mortar team nearby; this team will be referred to as the less effective team. The FSTs chosen for analysis were identified by Subject Matter Experts (SMEs), trainers at the facility, who stated that the two FSTs' completion of the mission were representative of more effective and less effective teams.

It must be stressed at this stage that the teams analysed were in the initial stages of pre-deployment training. They return to the training facility again after an additional number of months training together before they are deployed. The actions they make cannot be judged against deployment standards and instead must be couched at the preliminary training level.

Data Collection

The computer system at the training institution at which the case study occurred automatically collects a series of data. All voice communications that occurred within the scenario were recorded, along with a simulated video representation of all actions that were undertaken. In addition to this, a series of conversations with SMEs involved in the case study were voice recorded, providing a full account of each mission. The communication acts that occurred within each team were transcribed using a laptop computer and the Microsoft Excel computer software package. The conversations with SMEs were also transcribed in the same manner and a series of notes were taken from the simulated video representations, which were also transcribed.

Data Reduction and Analysis

In order to explore the decision-making process undertaken by the two FSTs, the EAST methodology (Stanton, Baber and Harris 2008) and coding-based modelling were employed. EAST analysis enables the identification of the core factors involved, and the coding provides insight into the interaction between these factors.

EAST

EAST contains a combination of Human Factors methods designed to systemically explore a scenario. Previous applications of the methodology have included insights into decision-making within a number of complex decision-making domains, including Brigade and Battle Group military teams (Stanton, Baber and Harris 2008); within Air Traffic Control teams (Walker, Stanton, Baber et al. 2010); and Police and Ambulance emergency response (Houghton, Baber, McMaster et al. 2006).

Population of the F3 Model through Coding

In order to identify the links between the factors, a coding scheme was applied to the communication transcripts of the more effective and the less effective teams. This coding scheme was developed in Chapter 1 based on a grounded theory-based approach. The core factors affecting team performance were drawn from the literature and developed into coding categories along with a series of coding rules presented earlier in the book. Below, two examples are highlighted to illustrate the manner in which the links are coded:

> Hello FST2 this is Apache have my sights on the area of grid ** I have a technical vehicle in a compound ready to copy grid.

This first extract is taken from the more effective team's transcript and has been labelled as representing the link *Coordination – Situation Awareness*. The Apache pilot's coordination with the Fire Support Team through their sighting enabled the Fire Support Team to hold correct Situation Awareness regarding enemy assets in their area of operations.

> JFC: FST roger he wasn't aware that you were about to engage with guns, info that should have been passed up by your JTAC, therefore he needs to be told about the deconfliction before you continue over.

This second extract is taken from the less effective team's transcript and has been labelled as representing two breakdowns: a breakdown in the link between *Coordination – Situation Awareness –* due to a lack of coordination between the friendly call signs involved in the mission, the defective crew have a poor SA regarding engagements occurring very close to their location. Additionally, it represents a breakdown between the link *Coordination – Schema*: due to an incorrect schema of deconfliction procedures, friendly assets are not able to coordinate their engagements effectively.

Inter-rater Reliability

Following the procedure of earlier chapters the author's coding of the communication transcripts, both for communication acts and for links within the F3 model, and for the CDA, were subjected to inter-rater reliability analysis. A coding scheme was developed for each of the three coding analyses, providing rules for each coding category alongside an example for each rule. Explanations were also developed for each task step in the CDA and for each communication act within the communication transcripts to ensure that the independent raters were provided with a clear account of any technical language present. At this stage, two independent raters were asked to code 10 per cent sections for each of the three coding analyses. The responses of the independent raters were recorded on response sheets and then entered into a Microsoft Excel software package on a laptop computer. The Microsoft Excel software package was then utilised to calculate a percentage agreement between the two independent raters and the author's initial analysis. The results of this analysis are presented below in the relevant Findings sections.

EAST Findings

This section presents the results of the analysis first exploring the EAST analysis before turning to the coding analysis.

Cooperation and Coordination

Hierarchical Task Analysis (HTA)   The complete HTA for the less effective and more effective teams are too large to be presented in full. The overall task structure of the two teams is similar, as shown in Figure 6.2.

Despite the same overarching goals, the less effective team has a far higher level of tasks, 1,571 compared to 1,095 in the more effective team. From this it could be hypothesised that the less effective teams required a greater number of tasks in order to accomplish the same number of goals. It appears that the more effective team was more efficient, using fewer task steps to accomplish the mission successfully. This finding is in line with previous research, such as that by Orasanu (2005) who stated that 'managed task loads' were indicative of effective team performance, and aligns with the results of the previous chapter.

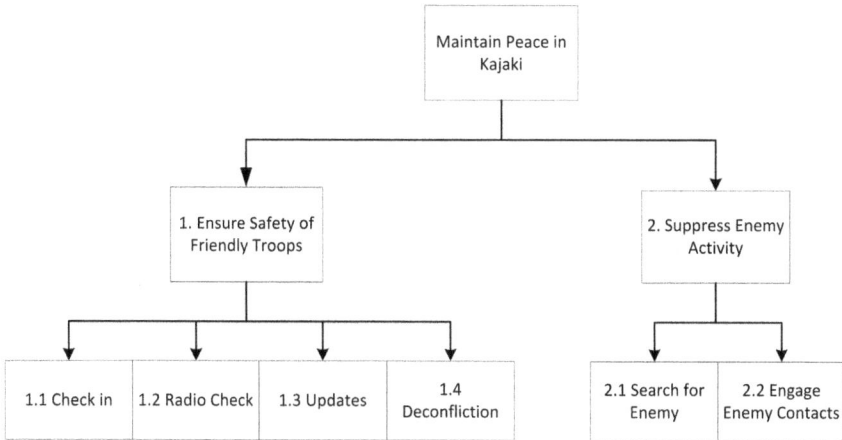

Figure 6.2   Goal structure

The initial sub-goal of *Check in* provides an example of the efficient manner in which the more effective team completed sub-goals and the issues that arose in the less effective team's goal completion. In the *Check in* sub-goal the more effective team conducted 84 tasks compared with 119 by the less effective team, despite both teams having the same number of assets to check in. This increased number of task steps within the less effective team can be explained by a number of breakdowns. An example of these breakdowns can be found in the *Check in* between Apache and the Joint Fires Cell, which involved confusion over which radio network the Apache should be using for the *Check in*. The Apache attempted to *Check in* on an incorrect radio network, received no response and was eventually informed by a Fire Support Team of the correct radio network they should be using for the *Check in* procedure. This *Check in* was also hampered later on by a further communication problem that led to a delay in the *Check in* procedure. There were no such problems with the *Check in* procedures performed by the more effective team, with all *Check in* procedures running smoothly with concise transmission of relevant information.

A further example of the problems associated with task completion in the less effective team is illustrated in the sub-goal of *deconfliction*. Deconfliction can be defined as an activity 'to reduce the risk of collision in (a combat situation, airspace, etc.) by separating the flight paths of one's own aircraft or airborne weaponry. Also: to coordinate (one's aircraft) in this manner' (Oxford English Dictionary 2010). Within the sub-goal of *deconfliction* a greater number of tasks were undertaken by the less effective team: 339, compared to 144 in the more effective team. This is not due to a greater level of *deconfliction* being required within the less effective team; rather, it is due to a number of issues associated with the *deconfliction* process. Three main issues arose within the less effective team. Firstly there was an incorrect understanding of the *deconfliction* procedures which led to discussions over the correct procedures during mission performance. Due to this incorrect understanding of the procedures, *deconfliction* information was not sent in a timely manner to those who needed it, resulting in an urgent *deconfliction* message when assets were already firing over a Fire Support Team's location without informing that Fire Support Team. This confusion also led to a great deal of confirmation of *deconfliction*, with assets questioning and also double-checking one another's *deconfliction*. This increased the *deconfliction* tasks and also delayed engagements of enemy assets.

The efficient task completion illustrated by the more effective team represents a coordinated environment where each member is capable of passing the correct information to other agents. This finding is consistent with the theory of DSA, which posits that information regarding the requirements of team members is distributed across the system and team members are able to access this information through interacting with the system (the idea of transactive SA) (Stanton, Salmon, Walker and Jenkins 2009a).

In addition to this, the more effective team ensured that others received information related to their specific tasks, for example, deconfliction information, and trusted one another to do this, thus demonstrating a level of cooperation. In the less effective FST low levels of cooperation were illustrated, for example, through the double-checking of one another's deconfliction, illustrating a lack of trust in team members' ability.

Coordination

Coordination Demands Analysis (CDA)   The summarised results of the CDA analysis are presented below in Table 6.1. The CDA undertaken in this chapter was subject to inter-rater reliability tests following the procedure outlined above under 'Method'. The results of the reliability testing revealed a mean agreement of 90 per cent with the initial CDA analysis conducted.

The CDA reveals that although the less effective team conducted a greater number of teamwork tasks in a direct numerical comparison with the more effective team, a percentage comparison (based upon the percentage of teamwork tasks in relation to the overall number of tasks undertaken by each team) reveals a similar level of teamwork tasks between the two teams (91 per cent in the less effective team and 90 per cent in the more effective team). Analysis of the team tasks revealed a higher level of coordinated activity within the team tasks in the

Table 6.1   CDA results for the more effective and the less effective teams

	Less effective	More effective
Total task steps	1250	833
Total individual work	110 (9%)	85 (10%)
Total teamwork	1140 (91%)	748 (90%)
Mean Total Co-ordination	2.67 (86%)	2.83 (94%)

more effective team, 94 per cent, than in the team tasks in the less effective team, 86 per cent, suggesting that a greater quality of coordinated activity occurred within the more effective team tasks compared to the less effective team.

Communication

Communication Usage Diagram (CUD)   As with the research presented in Chapter 5, the results of the CUD analysis, illustrated below in Figure 6.3, reveal that there is a greater level of communication within the less effective team compared to the more effective team.

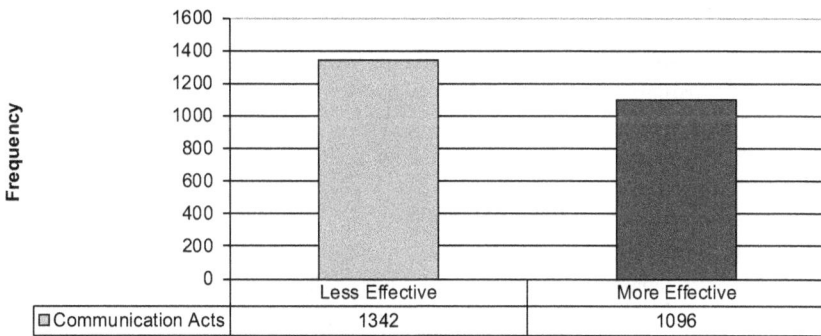

	Less Effective	More Effective
Communication Acts	1342	1096

Figure 6.3   CUD analysis for the more effective and the less effective teams

Radio networks   In order to explore the communications further, an analysis was undertaken of the radio networks utilised during the mission for both the more effective and the less effective crew. Figure 6.4 below represents a summary

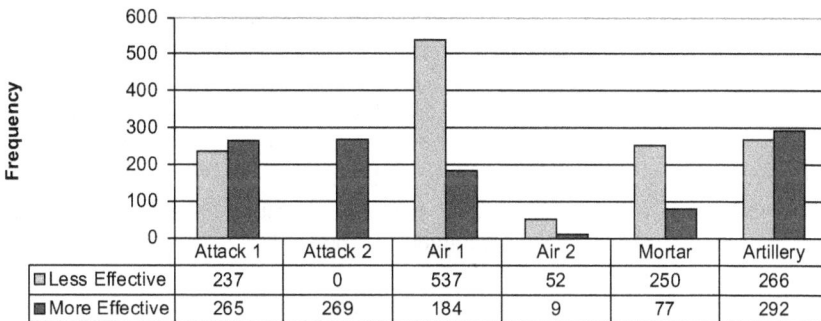

	Attack 1	Attack 2	Air 1	Air 2	Mortar	Artillery
Less Effective	237	0	537	52	250	266
More Effective	265	269	184	9	77	292

Figure 6.4   Radio networks utilised in the more effective and the less effective teams

of this analysis, showing the radio networks used by the teams, as well as their frequency of use.

Within the more effective team there was one additional radio network utilised (Attack 2) which, although present, was not used by the less effective team. This represents an increase in the communication channels exploited in the more effective team compared with the less effective team. A summary of the uses to which the radio networks were assigned is as follows:

- Attack 1 is used for FSTs to talk to Apache and Jets in the less effective team, and is used for FST2 to talk to Jet 2 in the more effective team.
- Attack 2 is not used in the less effective team, and is used for FST1 to talk to Jet 1 within the more effective team.
- Air 1 is used for JFC to talk to air assets and FSTs by both teams.
- Air 2 is used for the jets to talk to one another as an internal air network.
- The Artillery net is used by both teams for the FSTs to talk to guns and GMLRS.
- The Mortar net is used by both teams for mortar teams to talk internally to one another.

This illustrates that the more effective team split its two attack elements and put them onto separate radio networks, whereas the less effective team kept all attack assets on the same radio network. These options both have advantages and disadvantages associated with them. If the networks are busy it makes sense to add additional networks to ensure all messages can be sent. On the other hand, increasing the number of radio networks further distributes the team and allows the opportunity for information to be lost or not passed to the correct people. In this mission the splitting of the two attack assets on to different networks worked well, as both attack assets were also using Air 1 and Air 2 radio networks. Examination of the distribution of communications across the networks reveals that there was a greater frequency of communication occurring within the Attack networks in the more effective team; more than double that of the less effective team. From this it could be hypothesised that the more effective team needed to utilise a second attack network in order to ensure the effective flow of information. In the less effective, team more than three times the amount of communication occurred within the Air networks compared with that produced by the more effective team. Although an additional Air network was present, this was only infrequently used by the less effective team. From this it could be hypothesised that the more effective team was more efficient at managing their use of the communication networks available in order to create a well balanced communication network.

Coding   In order to gain a more in-depth understanding of why there was a higher level of communication acts present in the less effective team, the type of communication act has been analysed. Through the use of open coding the data was divided into a number of categories that represent the context of each communication act. The following quotation is an example of an extract from the communication transcripts:

> Hello JFC this is FST2 just confirming your UAV asset is still eyes on the HMG in that vehicle at grid *** just confirm no civilians or friendly forces in that vicinity over.

This extract was taken from the more effective team and illustrates the Fire Support Team requesting confirmation that an asset has eyes on (can see) a Heavy Machine Gun (HMG) at a specific grid and if the JFC can confirm that that there are no friendly or civilian personnel in the area. This extract was coded as *request information*.

In order to illustrate the reliability of the communication transcript coding, inter-rater reliability testing was conducted following the procedure described above under 'Method'. An agreement of 80 per cent was derived from the comparison of the coding between analysts. Table 6.2 below illustrates the percentage of

Table 6.2   Coding analysis summary for the more effective and the less effective teams' transcripts

Code	Less effective n (%)	More effective n (%)
Acknowledge	224 (17%)	217 (20%)
Confirm	34 (3%)	27 (3%)
Radio Check	42 (3%)	17 (2%)
Read-Back	169 (13%)	54 (5%)
Ready	23 (2%)	15 (1%)
Repeat	19 (1%)	6 (0.5%)
Request	6 (1%)	6 (0.5%)
Request Acknowledgement	24 (2%)	23 (2%)
Request Asset	9 (1%)	1 (0.1%)
Request Confirm	2 (0.1%)	1 (0.1%)
Request Granted	48 (4%)	57 (5%)
Request Information	101 (8%)	135 (12%)
Request Read-Back	5 (0.3%)	4 (0.3%)
Request Ready	24 (2%)	15 (1%)
Request Repeat	13 (1%)	18 (2%)
Request To Speak	73 (5%)	77 (7%)
Request Task	0	1 (0.1%)
Send Information	437 (33%)	345 (31%)
Send Order	37 (3%)	37 (3.4%)
Wait Out	48 (4%)	38 (3.4%)
Wrong	4 (0.3%)	2 (0.2%)

communication acts that fell into each coding category in both the more effective and the less effective teams.

The more effective team had a higher level of *request information* codes than the less effective team (in relation to the total number of communication acts within each team). This could illustrate a more proactive approach to gaining relevant information in the more effective team; however, it could also illustrate an inability in the team members to appropriately send the correct information to team members who required it.

The less effective team contains a higher level of *read-back* codes than the more effective team (in relation to the total number of communication acts within that team). Again, this finding could be interpreted in a positive or a negative manner. It could be that there was a lower level of trust in the appropriate transfer of information within the less effective team, illustrating that the team members did not trust one another to receive the information being passed. However, it could also represent a more rigorous communication structure in which closed loop communication was enforced. Closed loop communication is a process in which information is passed to a receiver and the receiver then explicitly confirms that they have received this information (Salas, Sims and Burke 2005, Wilson et al. 2007). Previous research has emphasised the benefits of this communication strategy, arguing that it ensures that information is correctly transferred (Salas, Sims and Burke 2005, Wilson et al. 2007).

Communication and Cooperation

Social Network Analysis (SNA)   The results of the Social Network Analysis reveal some interesting findings regarding the communication structures of the more effective and the less effective teams. The social networks derived from the analysis are presented in Figure 6.5 and Figure 6.6.

The lines on the networks reveal communications between the two agents they link. The strength of the lines represents the frequency of that communication, with the numbers within the lines illustrating the frequency of communication acts. A strong bold line illustrates a high level frequency of communication between those two agents. A weaker line represents a lower level of communication frequency. Initial examination of these networks reveals a lower level of prominent (frequent) links within the more effective team, compared with a high level of prominent (frequent) links within the less effective team. In order to explore this difference further, a number of graph theory metrics were derived from the networks, including: cohesion; diameter; density; sociometric status; and centrality. The following is a summary of the results:

- **Cohesion** represents the level of cohesive, mutual connections between agents. Figure 6.7 depicts the cohesion values for the more effective and the less effective teams. The results reveal a higher level of cohesion in the less effective team.

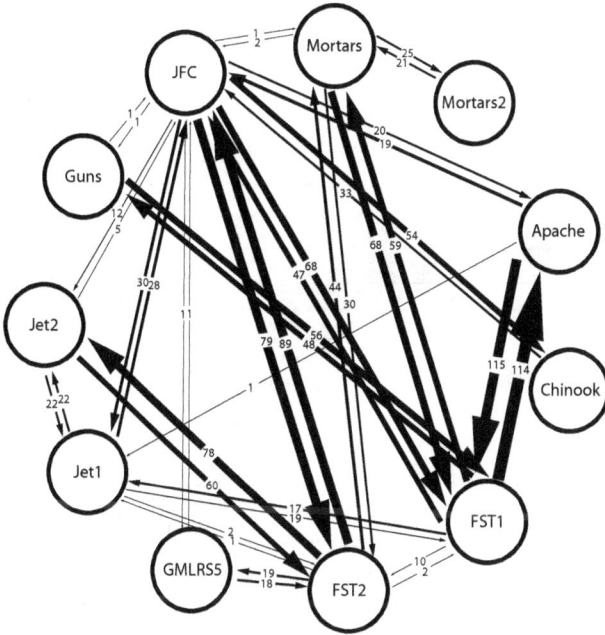

Figure 6.5   Less effective team social network

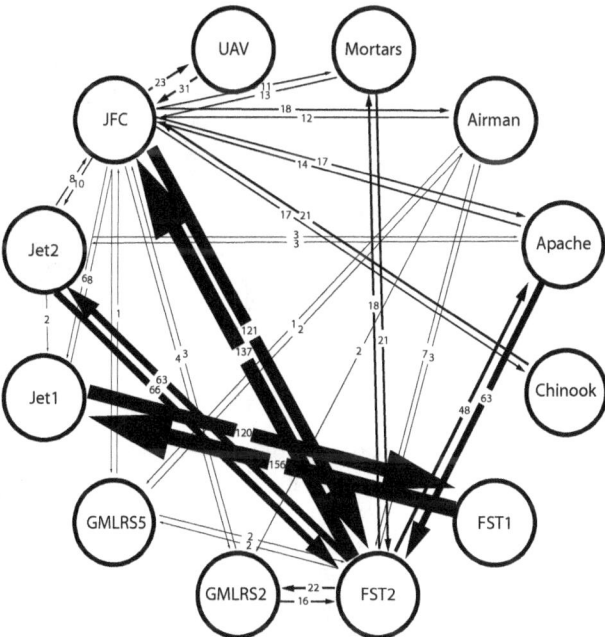

Figure 6.6   More effective team social network

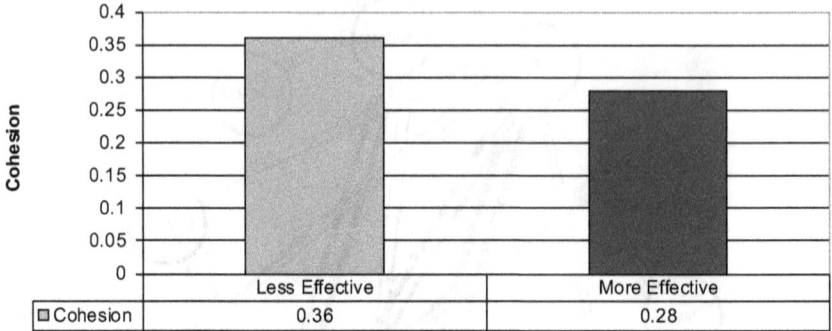

Figure 6.7   Level of cohesion in each team

- **Density** can be defined as the total number of interconnections within a network in relation to the total number of possible interconnections (Walker, Stanton, Kazi et al. 2009, Walker, Stanton, Salmon, Jenkins, Rafferty and Ladva 2010). Figure 6.8 below depicts the density levels for the more effective and the less effective teams. The results reveal that the less effective team contains a higher level of density than the more effective team.

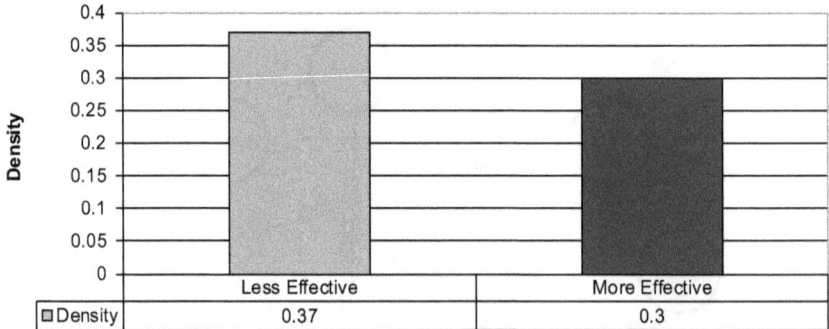

Figure 6.8   Level of density in each team

- **Sociometric Status** is a measure of the number and frequency of in and out communication links an actor has in relation to other actors in the network (Houghton, Baber, McMaster et al. 2006). It is a metric that describes how prominent that actor is as a communicator. Sociometric status thus represents how much an individual communicates (Walker, Stanton, Kazi et al. 2009, Walker, Stanton, Salmon, Jenkins, Rafferty and Ladva 2010). Table 6.3 summarises the sociometric status statistics. There is a greater level of sociometric status within the less effective team than the more effective team, with respect to both the sum and the mean value.

Table 6.3   Sociometric status statistics for each team

	Less effective	More effective
Minimum	3.9	0.818
Maximum	62.3	53.545
Sum	268.4	199.273
Mean	24.4	16.606
Variance	343.642	270.733
Standard Deviance	18.538	16.454

- **Centrality** represents the distance between an individual and other individuals within the network (Houghton, Baber, McMaster et al. 2006). A high level of centrality represents a high potential number of people an agent could talk to. Table 6.4 below summarises the centrality statistics. A greater level of centrality is revealed in the more effective team compared with the less effective team, with respect to both the total amount and the average amount of centrality.

Table 6.4   Centrality statistics for each team

	Less effective	More effective
Minimum	3.82	4.085
Maximum	8.682	10.042
Sum	63.197	74.979
Mean	5.745	6.248
Variance	1.564	1.868
Standard Deviance	1.251	1.367

The SNA has revealed higher levels of sociometric status, cohesion and density in the less effective team. In contrast to this, the more effective team has higher levels of centrality. The levels of diameter in the social networks were identical (3) for the more effective and the less effective team. A higher level of sociometric status, cohesion and density within the less effective team represents a greater number of connections between agents – both formally and informally, in addition to a high proportion of all possible connections being used. Although the more effective team had a communication network with close links between agents, shown through the centrality, it appears that they did not utilise the communication network to the same extent as the less effective team. In summary, the less effective team contained a strong, well-linked communication structure which was used to a greater extent than the communication structure of the more effective team.

In Chapter 5, the research surrounding the link between communication efficiency and effective performance was explored. Svensson and Andersson (2006) argued that effective performance is correlated with high levels of communication. Within this book it is suggested that the issue is more complex than the frequency of communication. Salas, Sims and Burke (2005) cited research by Roby (1968), which suggests that effective communication within teams involves ensuring that the right information is passed to the relevant person at the appropriate time. Continuing the notion that effective communication involves more than the frequency of communication, Bowers, Urban and Morgan (1992) state that high levels of communication are not always correlated with high levels of information transfers: they cite a study by Klienman and Serfaty (1989) in which teams in high-pressure situations were able to communicate the same amount of information in fewer communication acts.

Interpreting this finding in line with the DSA perspective, the lower level of communication between agents could be said to be indicative of a high level of task and role diversity. Within this large distributed team not all agents needed to communicate with one another, not all agents need to know all information. So although the more effective team had a high level of links between agents, they did not utilise these links to the same extent as the less effective team. This suggests that the more effective team were better able to communicate efficiently with those agents with whom they required communication, whereas in the less effective team, due to the diversity of roles a greater level of communication was passed where perhaps it was not necessary. The lower level of prominent links illustrates a more distributed communication structure showing that an 'all to all' network is not necessarily beneficial. Due to the high level of task diversity, the low coupling of tasks, not everyone needs to communicate with everyone else. The density metric emphasises this point – suggesting that a high proportion of links between agents is not necessarily beneficial.

Research conducted by Bryant (2006) supports the notion that different team members, specifically in the military, require different information. Bryant states that team members require 'unique perspectives on the same battle space to perform their specific roles' (2006: 190), due to their:

> different functions and responsibilities, different levels of command need different views on the underlying situation. Each unique view represents not just different levels of awareness but qualitatively different understandings by individuals with specific functional requirements within the C2 structure. (Bryant 2006: 200)

Situation Awareness and Schemata

Leximancer   In order to further explore the notion of DSA, the SA present in both the less effective and the more effective teams were studied in more detail. As in Chapter 5, Leximancer was used to derive information networks, as shown below in Figure 6.9 and Figure 6.10. The size of the nodes represents the elements' relevance to the network, that is, how frequently they occur and their relation to other elements within the network (Leximancer 2009). The links between the elements reveal the connectivity between the information elements (Leximancer 2009).

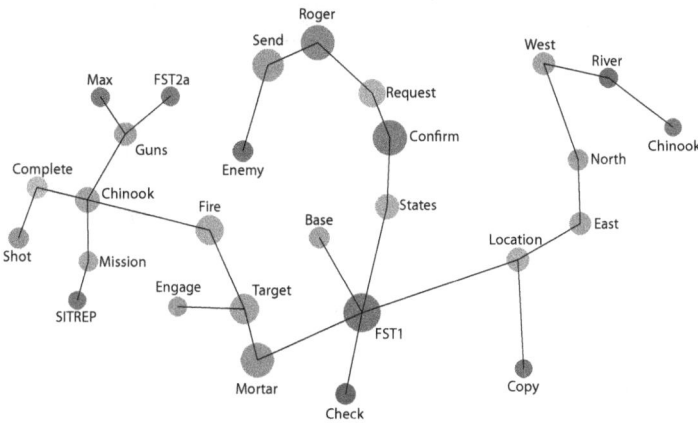

Figure 6.9   Less effective team information network

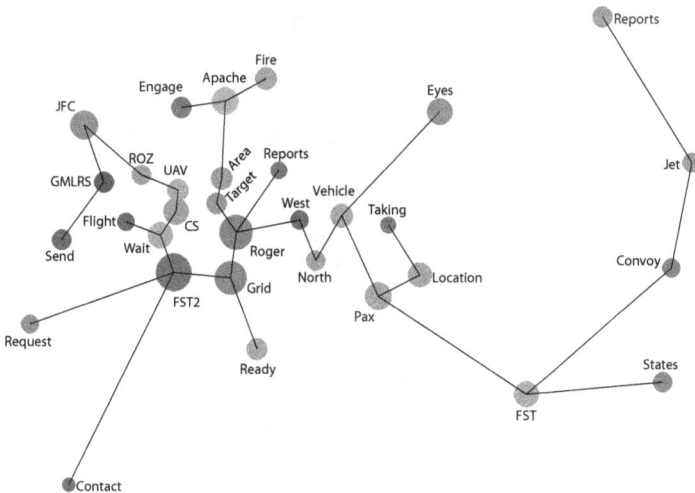

Figure 6.10   More effective team information network

Visual examination of the networks illustrates a high degree of variance between the two teams. Each team contains a number of different elements and the organisation of the elements in the two teams also differs substantially. The frequency and relevance of the elements are presented below in Table 6.5. Shaded elements are those that are present in both teams.

Table 6.5   Elements and relevance for the more effective and the less effective teams

	Less Effective			More effective	
Element	Count	Relevance	Element	Count	Relevance
request	118	100%	jet	120	100%
FST1	118	100%	FST2	117	98%
confirm	111	94%	eyes	111	92%
roger	106	90%	grid	103	86%
send	93	79%	JFC	102	85%
target	86	73%	states	96	80%
fire	81	69%	FST 1	80	67%
location	72	61%	pax	74	62%
states	64	54%	reports	71	59%
west	53	45%	fire	66	55%
east	52	44%	roger	65	54%
base	50	42%	apache	65	54%
rounds	48	41%	convoy	58	48%
north	45	38%	request	53	44%
guns	43	36%	vehicle	44	37%
mission	42	36%	location	43	36%
complete	40	34%	ready	39	32%
engage	37	31%	UAV	38	32%
shot	36	31%	wait	38	32%
FST2	35	30%	area	35	29%
SITREP	35	30%	send	35	29%
river	33	28%	call sign	32	27%
check	32	27%	engage	31	26%
copy	32	27%	ROZ	29	24%
enemy	28	24%	target	28	23%
Chinook	24	20%	north	28	23%
max	16	14%	GMLRS	25	21%
			clear	24	20%
			west	22	18%
			contact	22	18%
			flight	14	12%
			taking	13	11%

Exploration of the elements reveals that there are differences between the elements present in the less effective and the more effective teams. In the less effective team, 55 per cent of the elements are common to both teams, whereas within the more effective team only 37.5 per cent are common to both teams. From this it could be suggested that the more effective team utilised a high level of information – 62.5 per cent – that the less effective team did not utilise. In addition to this, and more importantly, the elements are connected in different ways. As in the previous chapters, graph theory metrics were derived from the Leximancer networks. The results of this analysis are presented below in Table 6.6.

Table 6.6   Information network metrics

SNA Metric	Less effective	More effective
Density	0.034	0.031
Mean Sociometric Status	0.068	0.063
Mean Centrality	22	29
Diameter	12	17
Cohesion	0	0

The centrality metric is higher in the more effective team than in the less effective team, suggesting that the more effective team holds a greater level of more prominent information elements. The less effective team contains a lower level of diameter, suggesting that the information present within the less effective team is more closely integrated than it is in the more effective team.

The less effective team information network contained a lower level of diameter, suggesting (as was found in Chapter 5) that paths between elements were smaller and information was less distributed than within the more effective team. This link between diameter and integration, as raised by Walker, Stanton and Salmon (2011), was outlined earlier in Chapter 4 and Chapter 5. This finding is in line with that of Chapter 5 and can be explained by the theory of DSA.

In light of DSA, it could be suggested that a less distributed information network is inappropriate. Team members require different elements of information due to their diverse roles and goals, and interpret these elements of information in distinctly different ways, due to individual schemata (Stanton, Salmon, Walker and Jenkins 2009a, 2009b, Stanton, Stewart et al. 2006, Salmon, Stanton, Walker and Jenkins 2009). From this perspective there is no 'shared' integration of information and therefore it would not be expected that information at this level be tightly integrated. Indeed, the DSA approach would suggest that a system in which information was 'shared' and integrated could be detrimental to effective team performance (Salmon, Stanton, Walker and Jenkins 2009). Shared SA is considered to be almost impossible, due to unique schemata, and unnecessary due to the diverse range of roles and goals within a system (Stanton, Salmon, Walker and Jenkins 2009a, 2009b). Stanton et al. (2009a) describe DSA as heterogeneous, rather than homogeneous, arguing that this is a beneficial quality for DSA within large teams of teams where the aim is to have a large number of people working on a diverse range of tasks.

It is hypothesised within this research that the higher path lengths found in the more effective team are illustrative of the high level of role diversity in the team. The individual team members did not require tightly integrated SA of the entire system; rather, they needed to be able to appropriately extract the relevant information for them: DSA. As information is organised and integrated based upon underlying schemata, it is unsurprising that the path lengths between information at a systems level are high. The information integration would be different for each agent based upon individual schemata and role.

Examination of the networks from the DSA perspective provides further weight for the argument raised earlier that 'good' SA is not everyone knowing everything. Within the more effective team there is a lower level of integration between information elements. This higher level of path lengths between elements is indicative of individuals 'picking and choosing' those information elements that are relevant or required for their role. Within the less effective team, the greater integration of elements suggests that there was a greater likelihood of all agents attending to all elements whether they were appropriate or not. The results of this research are in line with work by Cooke and Gorman, who support the notion of 'very specialised teams having little cognitive overlap among members and less specialised teams having more cognitive redundancy' (Cooke and Gorman 2006: 271).

Research by Atkinson and Moffat (2005) provides further weight to the argument that a lower level of integration could be beneficial to the information network within a team of members with diverse roles and responsibilities. Atkinson and Moffat argue that:

> Information is then taken by an individual and given meaning within their individual context ... each person will still have a different perspective on the key issues. (2005: 90)

The distribution of elements is important as it enables insights into the manner in which each element is interpreted and integrated. This distribution represents shorter pathways between each information element (Walker, Stanton, Salmon, Jenkins, Rafferty and Ladva 2010). The lower level of distribution is only beneficial if the integration is appropriate, as is illustrated in Chapter 5. In order to explore these connections further, thematic analysis of the networks was undertaken.

The thematic analysis allows further insights into the manner in which information is connected and the way in which this affects its interpretation and the development of SA. The thematic maps derived for the more effective and the less effective teams are presented below in Figure 6.11 and Figure 6.12, and explored in more depth in Table 6.7 on the following page.

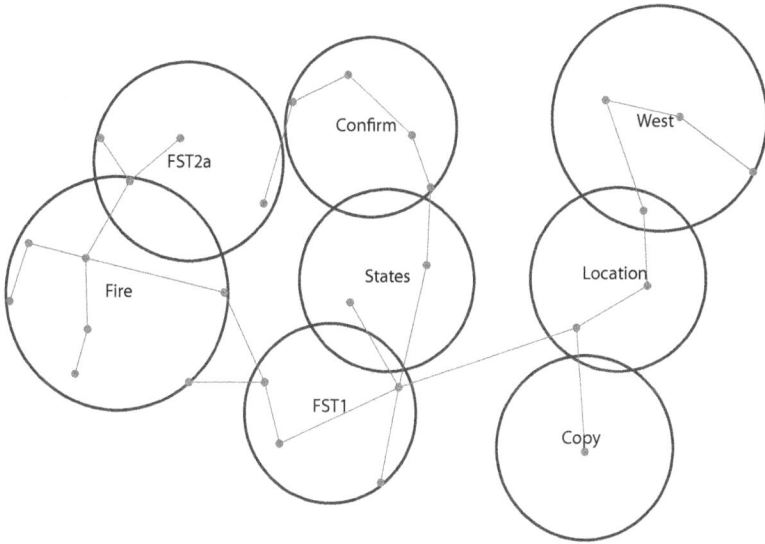

Figure 6.11 Less effective team themes

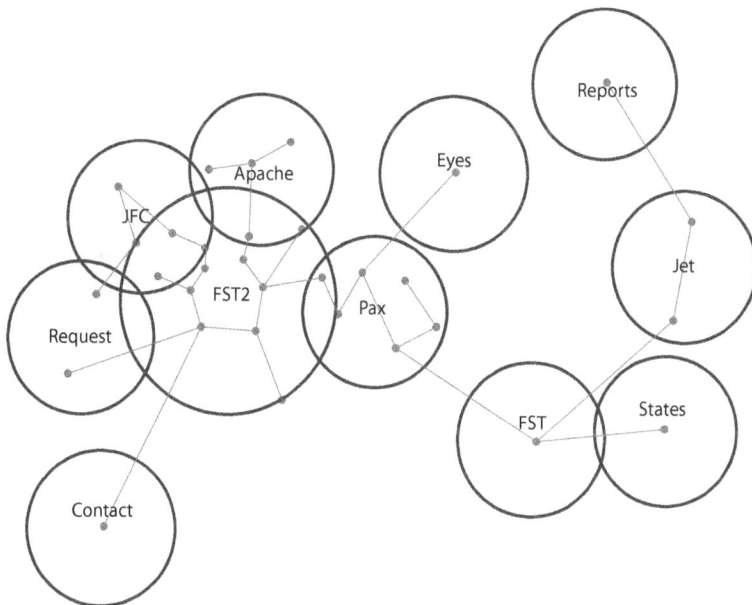

Figure 6.12 More effective team themes

Table 6.7   Core themes and connectivity for the less effective and the more effective teams

Less effective		More effective	
Theme	Connectivity	Theme	Connectivity
fire	100%	FST2	100%
confirm	96%	grid	56%
FST1	84%	pax	45%
location	40%	jet	12%
states	30%	eyes	11%
west	25%	request	10%
FST 2	22%	FST 1	8%
copy	7%	reports	5%
		states	5%
		contact	2%

The thematic groupings within the elements represent genotype schemata. In the less effective team the most connected theme is *fire*, indicating an emphasis on engagements in this team, whereas for the more effective team the most connected element is *FST2* illustrating the emphasis on friendly units.

Within the more effective team the themes appear to be focused around the transfer of information with themes including *reports* and *requests*. In addition to this, the information focus is on a high fidelity of information illustrated by the themes of *grid* and *eyes* – eyes referring to an asset being 'eyes on' a target or area, which means that the target or area is being directly observed. In the less effective team the theme of *west* represents a focus on a lower fidelity of information, stating that a target or area is *west* provides far less information than stating its *grid* reference.

The less effective team theme of *confirm* could initially appear to be positive, representing a focus on confirmation, which would be expected to be beneficial. However, closer examination of the transcripts and analysis reveals that the *confirm* theme is present after the incident of fratricide occurs – its relevance occurs when the FST are attempting to *confirm* with the Apache exactly what happened, as is illustrated in the transcript extract presented below ('pax' is a military term for 'person'):

Apache this is FST1 just confirm that was one single pax in the village.

The detail provided by Leximancer allows the identification of the exact areas of the transcripts from which the themes are drawn, which enables an accurate description of the context of each theme, providing valuable insights into the two teams' SA and schemata.

In light of the discussion of the key information elements, it appears that within the more effective team a schema focused on information transfer existed, whereas within the less effective team the dominant schema was that of engage. The two teams' completion of the mission highlight these schemata and the way in which they impacted on task performance. As has been discussed earlier, the way in which people interpret the environment and information is based upon their individual schemata (Stanton, Salmon, Walker and Jenkins 2009b). In this section of the book, the impact of inappropriate, non-compatible schemata on team performance has been explored. Research suggests that under conditions of stress people are impacted by a 'confirmatory bias' (Dean and Handley 2006, Famewo, Matthews and Lamoureux 2007, Greitzer and Andrews 2009) and as a result of this 'people dismiss inconsistent information when making a decision' (Famewo, Matthews and Lamoureux 2007: 32). It could be suggested that the less effective team was focused upon the *fire* schema and dismissed any information that was not in line with this schema, resulting in the incident of fratricide.

Summary

The results of the EAST analysis for the more and less effective teams are summarised below in Table 6.8.

Table 6.8 **Summary of the more effective and the less effective teams' EAST analysis**

		Less effective	More effective		
	HTA	Higher number of tasks	Lower number of tasks	Coop and Coord	
	HTA	Higher level of ineffective tasks	Lower level of ineffective tasks		
	CDA	Similar level of teamwork tasks	Similar level of teamwork tasks	Coord	
	CDA	Lower level of coordination – 86%	Higher level of Coordination – 94%		
EAST	CUD	Higher level of communication acts – 1342	Lower level of communication acts – 1096	Comms	F3 Model
	SNA	Slightly higher levels of density, cohesion and sociometric status	Slightly lower levels of density, cohesion and sociometric status	Comms and Coop	
	SNA	Similar levels of centrality	Similar levels of centrality		
	IN	Lower level of centrality	Higher level of centrality	SA and Schemata	
	IN	Lower diameter	Higher diameter		

Interaction of Factors

The coding of the communication transcripts in terms of the links between factors was subject to inter-rater reliability tests, as is outlined above under 'Method'. The analysis revealed a percentage agreement of over 85 per cent. The results of the coding for the transcripts of the more effective and the less effective teams are illustrated below in Figure 6.13.

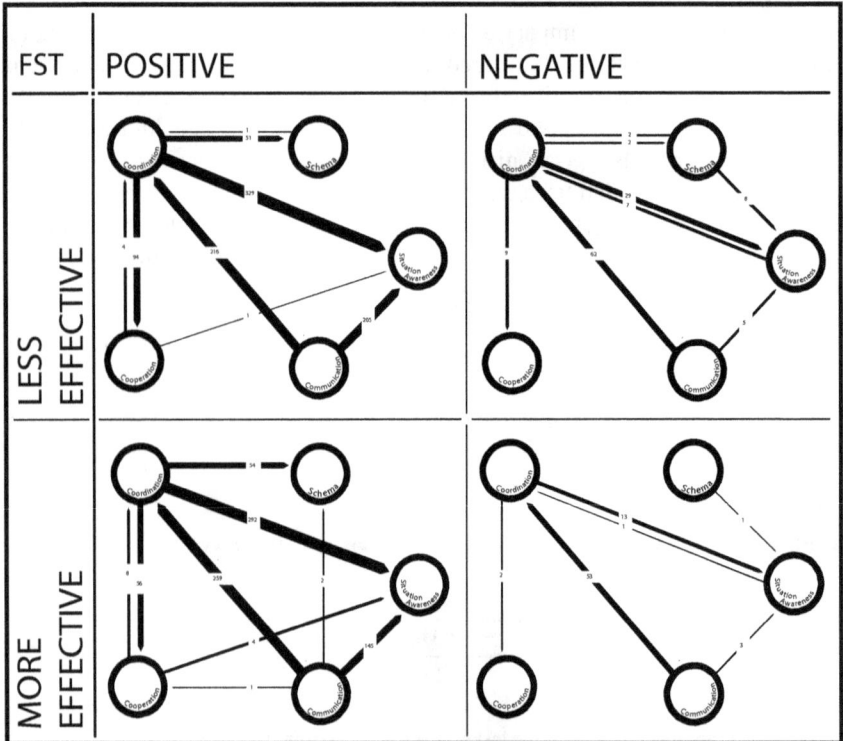

Figure 6.13   More effective and less effective teams' positive and negative links

Comparison of the populated models shown above in Figure 6.13 reveals differences in the frequency and presence of the links between the less effective and the more effective teams. These differences are explored further in Table 6.9 and Table 6.10.

A greater frequency of positive links is present in the less effective team model than in the more effective team model. Although this is counter-intuitive, it appears that this greater frequency of positive links is outweighed by the greater frequency of negative links within the less effective team. In line with Chapter 5, it appears

Table 6.9   Frequency of positive and negative links in both teams

Frequency	Positive	Negative	Difference
Less effective	901	124	877
More effective	821	73	748
Total	1722	197	

Table 6.10   Presence of positive and negative links in both teams

Presence	Positive	Negative	Difference
Less effective	8	8	0
More effective	9	6	3
Total	17	14	

that the negative links have a disproportionate impact on the team compared with the positive links.

Examination of the presence of links reveals the importance of three core links across all four models: Communication – Coordination; Coordination – Situation Awareness; and Communication – Coordination. The link between Communication and Coordination seems to be the most important with respect to negative links, and the links between Communication, Coordination and Situation Awareness appear to be the most important with respect to positive links. From this it can be posited that the links between Situation Awareness, Coordination and Communication represent key causal factors in decision-making within teams.

The importance of SA is further emphasised by the modelling of links between the factors. The link between Communication and Coordination is higher within the more effective team than in the less effective team. From this it could be suggested that the more effective team was better able to appropriately communicate only that information required for a coordinated effort, whereas in the less effective team it could be suggested that the distinction between which agents needed which information in the coordinated mission was less clear. This finding is in line with the theory of DSA, in which the need to understand which agents require which information is emphasised, labelled as meta-SA (Stanton, Stewart et al. 2006, Salmon, Stanton, Walker and Jenkins 2009, Stanton, Salmon, Walker and Jenkins 2009a, 2009b).

Conclusion

The comparison of a more effective and a less effective Fire Support Teams' completion of a Close Air Support mission enabled a number of comparisons to be drawn. A number of core factors and core interactions associated with effective decision-making, as well as with breakdowns in this effective decision-making process, were identified. EAST was utilised to explore core factors, illustrating a correlation between more effective performance and high coordination activity, low levels of communication, high levels of information integration, and high levels of cooperation. Coding-based modelling enabled the identification of three prominent relationships between the factors of Communication, Coordination and Situation Awareness, with Coordination acting as a mediating factor between Situation Awareness and Communication.

The results of this research support the assertion developed in previous chapters that efficient, succinct and pertinent communication and information transfer is beneficial within a multi-role team of teams context, supporting the theory of DSA (Stanton, Salmon, Walker and Jenkins 2009a, 2009b, Salmon, Stanton, Walker and Jenkins 2009). Within such an environment, the constraints placed on team members to coordinate with one another add additional workload demands in the time-sensitive situations in which they operate. In order to ensure that accurate decisions can be made, team members must be aware of the information others require, and the information they themselves require, in order to ensure that team members are not overwhelmed with a host of communications and information transfer, much of which may be extraneous.

In summary, the research presented in this chapter provides validation of the theory of DSA and its importance to effective team performance. The research also provides validation for the ability of EAST, in collaboration with coding of communication transcripts, to support the identification of factors involved in fratricide incidents and the way in which these factors interact with one another.

Chapter 7
Comparison of Populated Models

Introduction

Chapters 4, 5 and 6 have explored case study incidents of fratricide from British military training institutions. Each case study presented different contexts: from small tank crew teams in traditional training scenarios to large Joint Fires Cells conducting multi-force operations, ending with Fire Support Teams conducting Close Air Support tasks. The aim of the research was to explore the decision-making and team processes that are involved in these situations, rather than the context-dependent task properties. Despite the situational task differences there are core theoretical principles that can be drawn from the processes involved. These core principles are explored within this chapter.

This chapter provides a comparison of the models developed from the case study research presented within the book. The results of the EAST analysis of each case study are used to draw out key differences and similarities. The populated models drawn from the case studies using coding are compared and contrasted through the use of Social Network Analysis and graph theory metrics.

The three case studies presented within this research represent three different levels of analysis. The tank crew study represents the lowest level of analysis, the JFC case study represents the highest level of analysis, and the FST case study sits between the two. Figure 7.1, on th following page, illustrates the overall military command structure and where each case study sits within this.

The ability of the case studies to explore a number of systemic levels aligns with general systems theory, upon which this book is based (Reason 1990, Leveson 2001, 2002, Paries 2006, Snook 2000, Svedung and Rasmussen 2002, Hutchins 1995). Paries discusses the relationship between different systemic levels, emphasising the impact that actions at one level have upon other systemic levels:

> [E]mergence is then simply a broader form of relationship between micro and macro levels, in which properties at a higher level are both dependent on, and autonomous from, underlying processes at lower levels. (Paries 2006: 46)

In order to explore this emergence, the authors of this book have researched incidents of fratricide at three systemic levels. The results, as discussed in this chapter, have revealed a complex relationship between the systemic levels representing the non-linear interactions hypothesised to exist in fratricide in Chapter 1. The ability of the method, and the model to explore, and explain, a

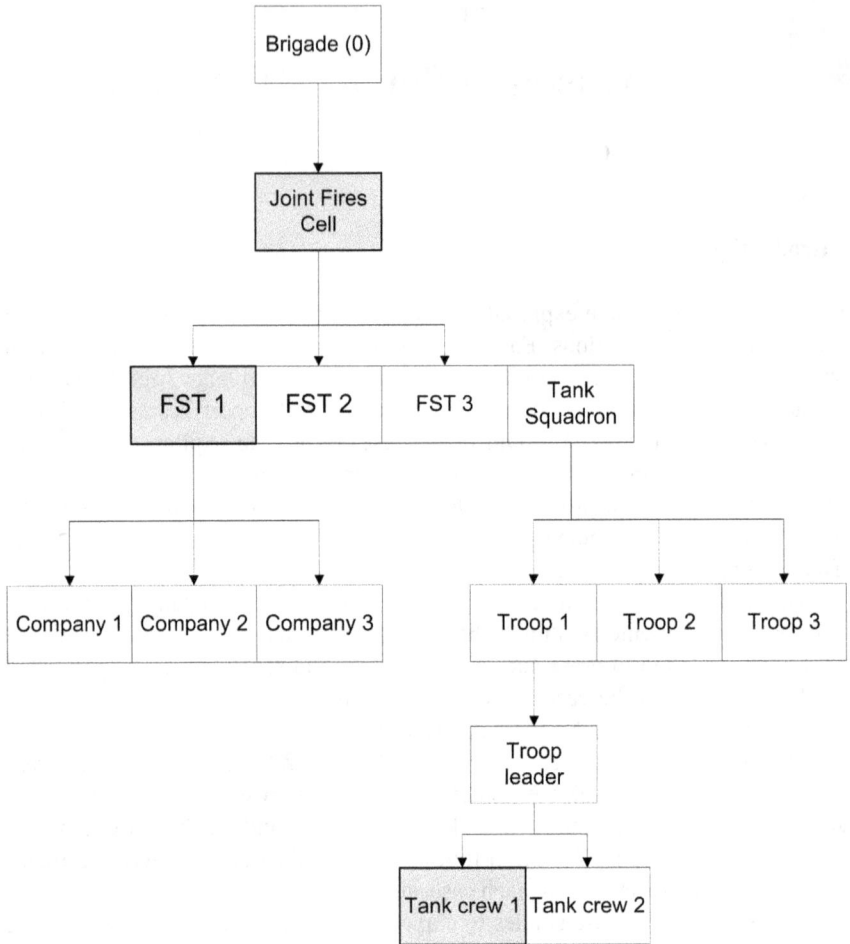

Figure 7.1   Systemic levels

number of systemic levels and a number of different units of analysis, provide a comprehensive analysis and support for the utility of the method and model.

Scenarios

Scenario One

The first case study undertaken involved the observation of British Army tank crews undergoing training within a simulated environment. A Battle Group was observed over the course of a week as it went through a series of training missions.

This research focused on one mission, a Combined Battle Group quick attack. In this mission the Battle Group was tasked with attacking and gaining control of a known enemy location. During the mission the members of one tank crew were identified by Subject Matter Experts as making effective decisions, appropriately engaging a number of enemy targets. Members of a second tank crew were identified by Subject Matter Experts as making less effective decisions, engaging a friendly reconnaissance vehicle under the mistaken belief that it was an enemy tank.

Scenario Two

A further case study was conducted observing Joint Fires Cells undergoing pre-deployment training in a series of connected high fidelity simulators. This training involved the RAF and the Army training together in order to coordinate ground and air attacks on enemy targets. The specific mission observed involved the force attending a meeting of village elders, alongside identifying and engaging a series of Improvised Explosive Device (IED) factories in the vicinity. Subject Matter Experts identified two teams, one representing an effective mission completion, effectively engaging all targets and attending the meeting; the other team failing to complete the mission effectively, having not engaged all targets and having mistakenly engaging a group of friendly British Special Force troops.

Scenario Three

The third case study focused on a Fire Support Team undergoing pre-deployment training within the context of a larger Joint Fires Cells structure. Within this scenario the Fire Support Team was undertaking a mission as though part of a wider Joint Fires Cell, but the focus of the training was on the Fire Support Team. In this example the teams were undertaking a scenario based on 'a normal day in Kajaki'. They were given no specific missions, but were tasked with the overall goal of maintaining peace within the area. During the mission a number of enemy targets engaged the teams. Subject Matter Experts again identified one team that effectively completed the mission, identifying and engaging appropriate targets; and a second team that was less successful, mistakenly tasking an Apache attack helicopter to engage their own location.

EAST Findings

Exploration of the results of the EAST analysis for each case study highlights a number of interesting findings. There appears to be a high correlation between the results of the two inter-team studies (JFC and FST), when compared to the intra-team study (the tank crew). The following part of the chapter presents the results, which are discussed in relation to the differences and similarities highlighted between the case studies.

Differences

Tasks   In the intra-team study (Scenario 1) a higher level of tasks was linked to more effective decision-making, whereas in the inter-team case studies a higher level of tasks was linked with less effective performance. From examination of the data it appears that the more effective team within the intra-team study achieved a greater number of mission goals, and therefore conducted a greater level of tasks. However, in the inter-team studies the more effective and the less effective teams both conducted the same number of mission goals, but the more effective teams were able to do this in a more succinct manner, requiring fewer task steps than the less effective teams. This finding is in line with previous research, such as that by Orasanu (2005) who stated that 'managed task loads' were indicative of effective team performance. The importance of the succinct nature of task work in inter-team studies may be further emphasised due to the additional resources required to deal with coordination overheads within these distributed environments (Fiore, Salas et al. 2003, Hess et al. 2000 cited in Fiore et al. 2003, Moffat 2003a, Espinosa, Lerch and Kraut 2002, Paris, Salas and Cannon-Bowers 2000). As is argued by Urban, Weaver, Bowers and Rhodenizer, team performance in distributed environments is more challenging and requires individuals to 'interact quickly to communicate critical information' (Urban, Weaver et al. 1996: 301).

Communication   Within the inter-team case studies a lower level of communication acts was associated with more effective decision-making, whereas in the intra-team case study a lower level of communication was linked to less effective performance. The research presented within this book has suggested a complicated relationship between communication and effective performance. Svensson and Andersson (2006) argue the impact of communications and communication frequency on team performance is complex; and this sentiment is echoed by MacMillan, Paley, Levchuk, Entin, Serfaty and Freeman, who hypothesise that 'Communication can be good or bad for team performance, depending on when it occurs and what else is going on at that time' (2002: 12).

The communication findings presented within this research are supported by the work of Espinosa, Lerch and Kraut (2002) who found that smaller teams utilise communication to a higher degree than larger teams. It could be suggested that between teams, less communication is beneficial in the context of the distributed environment; if the environment is made up of multiple teams and team members there are many more people to communicate with, yet there is only a certain level of communications that any individual can process due to our cognitive capacity (Salas, Sims and Burke 2005, Bowers, Urban and Morgan 1992, Bryant 2006). Within this environment it is important to be concise so as to not overload individuals (Hourizi and Johnson 2003, Moore et al. 2003). In line with the theory of DSA within the larger teams, the distributed nature of the environment means that many people do not require all of the information due to differing roles (Stanton, Stewart et al. 2006, Stanton, Salmon, Walker and Jenkins 2009a, 2009b, Salmon, Stanton, Walker and Jenkins 2009, Bryant 2006, Gorman, Cooke and Winner 2006), allowing communication of information to be more selective, whereas in the smaller, collocated, team there is a higher level of transactive elements requiring a greater level of communication. Paris, Salas and Cannon-Bowers (2000) support this suggestion; their research found that in teams in which there is a high degree of role diversity there is a lower level of communication than in teams where roles are tightly coupled and overlapping.

Higher levels of sociometric status were linked to more effective decision-making in the tank crew study, suggesting that a greater percentage of agents, making a greater contribution to the communication flow, is beneficial in line with the higher levels of transactive SA within this system. Conversely, in the inter-team case studies a lower level of sociometric status was linked with more effective decision-making, reaffirming the suggestion from DSA above that within inter-team scenarios lower communication levels are beneficial.

Situation Awareness The results of the Leximancer analysis revealed that there are differences in the information used by the less effective and the more effective teams. In addition to this, the level of integration between the elements, the connections between them and their relational pathways, also differ between less effective and the more effective teams. Across all three case studies there is a greater level of centrality in the more effective teams' information network than in the less effective teams' information network, as illustrated below in Figure 7.2.

Centrality is a metric which measures the connections between nodes in a network, how prominent nodes are (Walker, Stanton and Salmon 2011). Within an information network a high level of centrality is illustrative of a greater number of prominent information elements. This suggests that within the more effective

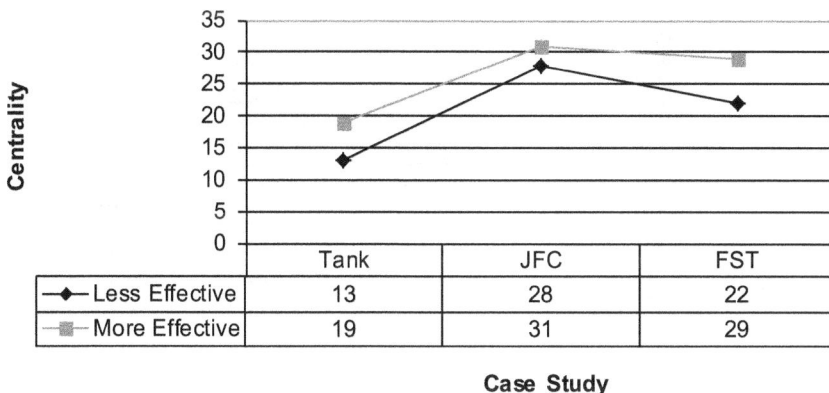

	Tank	JFC	FST
Less Effective	13	28	22
More Effective	19	31	29

Case Study

Figure 7.2   Centrality metric across all case studies

team, across all three case studies, information elements were more prominent. It could be hypothesised that centrality reflects an increased ability to identify and attend to important information within the more effective teams compared to the less effective teams, which were unable to discriminate between key and non-key information to the same degree – illustrated by the lower levels of centrality in the less effective teams.

In Chapter 5 and Chapter 6 the ability to identify and attend to appropriate information was discussed. In a large team, team members may be unable to process all available information due to a limited capacity to attend to information (Moffat 2003a, Boiney 2007, Bryant 2006, Dekker 2002, Woods et al. 1994, Neisser 1976), therefore team members would need to be able to decipher which information is appropriate. Research by Moore et al. (2003), Boiney (2007), Greitzer and Andrews (2009) and Bryant (2006) has emphasised the difficulties associated with identifying the correct information in the mass of information passed to decision-makers. Moore et al. (2003) and Hourizi and Johnson (2003) have highlighted the detriments of an inability to attend to the correct information, which can have consequences such as processing inappropriate information and not processing appropriate information required for developing SA.

The information networks also reveal differences between the case studies, specifically with respect to the level of diameter present in each information network, as is illustrated below in Figure 7.3.

Diameter is a metric which enables a measurement for the path lengths, or distance between nodes in a network. A high level of diameter is representative of a network where nodes have longer path lengths between them, are more distributed (Walker, Stanton and Salmon 2011). In relation to information networks, a high diameter value is illustrative of loosely integrated, more distributed information. Diameter is higher in the two larger, distributed, between-team case studies (JFC and FST), and lower in the smaller, collocated, within-team tank case study, for the more effective teams.

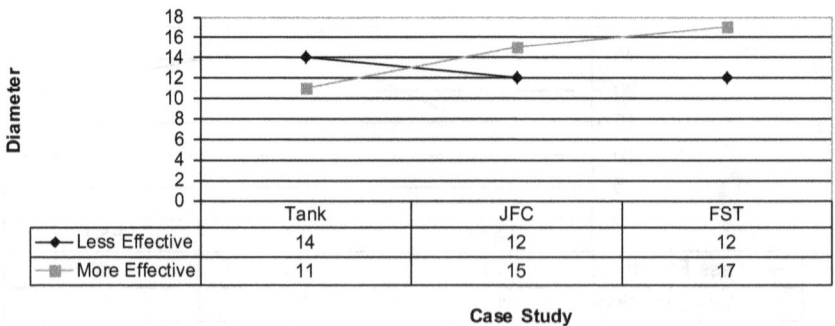

	Tank	JFC	FST
Less Effective	14	12	12
More Effective	11	15	17

Case Study

Figure 7.3   Diameter values across all case studies

It could be hypothesised from this result that within small, collocated teams with similar roles there needs to be a tightly integrated set of information elements with shorter pathways between elements of information (as is illustrated by the lower diameter level in the more effective team), but this does not, and should not, scale up to the larger, distributed between team (JFC and FST) systems, where everyone does not need to know everything (illustrated by the higher level of diameter in the more effective teams) and not all information is equally relevant to team members (Famewo, Matthews and Lamoureux 2007). To summarise: within teams you need quicker access to thematically organised information, but in larger between team systems there is a greater level of role diversity and correspondingly a greater diameter; across all of the more effective teams there is a greater level of more prominent information, but within larger teams of teams this information is more distributed, less integrated. Supporting this assertion is research by Cooke and Gorman (2006), which found that teams with highly specific roles require little cognitive overlap compared to teams who perform similar, or less specialised tasks. The results of this research illustrate the importance of information, but also the way in which this information is interpreted, with vignettes illustrating the misinterpretation of information across all of the less effective teams.

Similarities

Tasks Across all case studies a lower level of non-mission related tasks was considered beneficial to the decision-making process. Non-mission relevant tasks only serve to distract individuals and could overload their cognitive abilities unnecessarily. This finding is in line with previous research such as that discussed in Chapter 4 (Hutchins, Hocevar and Kemple 1999, Dismukes, Loukopoulos and Jobe 2001).

Coordination Levels of teamwork tasks were similar across all case studies in the more effective and the less effective teams. In each case study a higher level of coordination was found in the teamwork tasks of the more effective team. From this it may be suggested that across all case studies higher quality, rather than quantity, of teamwork was linked to effective decision-making.

Schemata A higher level of appropriate schemata is found in the more effective team for each case study. Schemata guide our interactive behaviour with the world (Neisser 1976). They direct individuals' attention and, therefore, their awareness of a situation and which aspects of the external environment to observe and process (Neisser 1976). This finding is in line with the theory of DSA, which is grounded in Neisser's Perceptual Cycle model, suggesting that appropriate schemata impact on the information individuals look for and attend to (Stanton, Salmon, Walker and Jenkins 2009b). Numerous other researchers have emphasised the role of past experiences and expectations on information interpretation and integration (Atkinson and Moffat 2005, Dekker 2002, 2005, Woods et al. 1994, Greitzer and

Andrews 2009, Famewo, Matthews and Lamoureux 2007, Dean and Handley 2006, Gadsen et al. 2008). Hirokawa, Gouran and Martz (1988), drawing on the available information from the Challenger space disaster, argue that mental constructs impact on the way in which information is attended to and utilised in the decision-making process, which may lead to people drawing the wrong conclusions

Summary

Each of the case study scenarios was analysed using EAST, and Table 7.1 below presents a summary of the results derived from the EAST analysis.

Table 7.1   Summary of all EAST results

		Less Effective			More Effective			
		Tank	JFC	FST	Tank	JFC	FST	
HTA	No. of tasks	—	+	+	+	—	—	Coop and Coord
HTA	Inappropriate tasks	—	+	+	+	—	—	
HTA	Crucial tasks	—	+	+	+	—	—	
CDA	No. of team tasks	/	/	/	/	/	/	Coord
CDA	Coord. level	—	—	—	+	+	+	
CUD	No. of comms	—	+	+	+	—	—	Comms
CUD	Non mission related comms	+	/	/	—	/	/	
SNA	Socio	—	+	+	+	—	—	Comms and Coop
SNA	Centrality	/	+	/	/	—	/	
SNA	Density	/	—	+	/	+	—	
SNA	Cohesion	/	—	+	/	+	—	
IN	Centrality	—	—	—	+	+	+	SA and Schema
IN	Diameter	—	+	+	+	—	—	

(Left margin label: EAST; Right margin label: F3 Model)

Note: / No difference; — Less; + More.

Across all studies, appropriate schemata – lower levels of non-mission relevant tasks and higher levels of coordination – are correlated to more effective decision-making. It appears that due to the high level of complexity, role diversity and difference in goals in inter-team missions, a more concise approach to task-work, communication and information transfer is beneficial, compared to the positive benefits of higher levels of communication and information transfer in intra-team situations. The finding of most importance from this case study is that high levels of communication and information are not always beneficial. When decision-making occurs within a diverse multi-force, multi-role context, not all agents need to communicate with one another and not all agents need to be aware of everything. Additionally, the importance of coordination is increased within this multi-team environment, mediating the relationship between communication and SA.The results of the EAST analysis can be explained using the theory of DSA (Stanton, Stewart et al. 2006; Stanton, Salmon, Walker and Jenkins 2009a, 2009b, Salmon, Stanton, Walker and Jenkins 2009). In the small tank crew team the tasks undertaken were tightly coupled with one another; they therefore required a great deal of transactive information elements. In light of this and the same geographical location and environment in which the crew were working, the team members held a large amount of transactive SA requirements. This high level of coupling meant that high levels of communication and information transfer were beneficial in this environment. as the likelihood of transferring irrelevant information was low.

Within the larger JFC and FST studies the tasks were loosely coupled, and team members were from both the RAF and the Army and had a diverse range of roles. This role diversity and the geographical spread of team members meant that the majority of information elements were different and, therefore, a greater frequency of information transfer and communication was not beneficial. From this it could be argued that within teams a greater level of communication is beneficial, whereas between teams a lower level of communication is beneficial.

Interaction of Factors

Although the results of the EAST analysis provide interesting insights into the causal factors associated with fratricide incidents, this book is concerned with the interaction of these factors and their resulting evolution into fratricide. In order to explore the interactions of the factors, the case study data has been inputted into the F3 model framework using coding of the communication transcripts. In this section, graph theory metrics are derived for each of the models, allowing for a statistical evaluation of their similarities and differences.

Model from Literature

Figure 7.4 represents the model derived from the literature as discussed earlier in Chapte 1. This model represents the most prominent connections as discussed by the literature (see Chapter 1); this refers to both positive links and negative links. The link with the greatest frequency is that between Communication and Situation Awareness. This link is closely followed in importance by the links between Coordination and Schemata and between Schemata and Situation Awareness. The links between Coordination and Communication, Coordination and Situation Awareness, and Schemata and Communication also represent prominent links. All other links have a frequency value of less than 10 per cent.

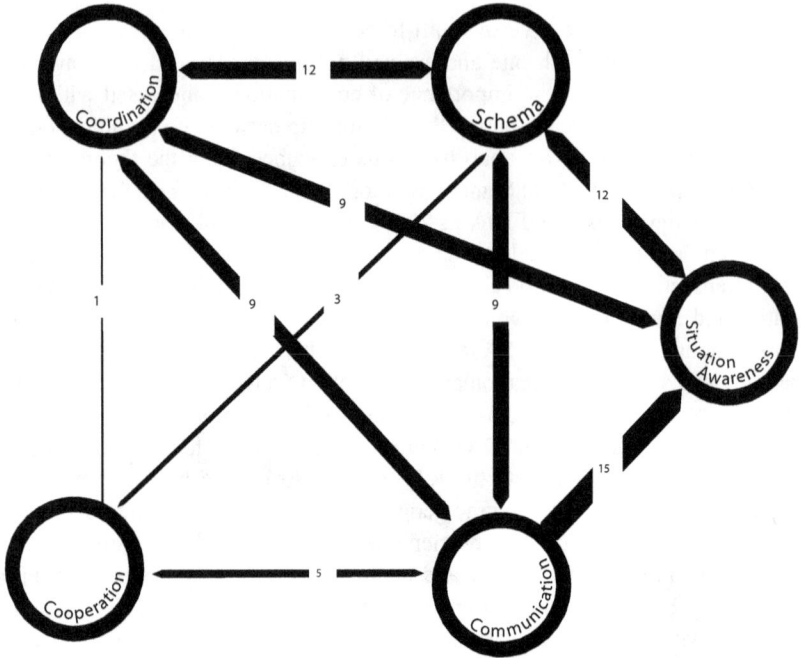

Figure 7.4   Model populated with data from the literature

Tank Model

Figure 7.5 represents the models derived from the tank crew case study as illustrated in Chapter 4. The four models represent the positive and negative links for the more effective and the less effective tank crews observed. In summary, the link between Situation Awareness and Communication seems to be the principal link for the less effective, fratricide, scenarios; in the more effective crew positive links between these two factors were higher, and breakdowns far lower. In the less effective tank crew, positive links were lower and broken links far higher. From this it can be posited that the link between Situation Awareness and Communication may be a key causal factor in fratricide. The model also allows identification of the role of Cooperation in the more effective tank crew – with its links to Communication and Coordination. In contrast, links to or from Cooperation were completely absent from the less effective tank crew.

JFC Model

Figure 7.6 presents the model populated with data from the JFC study as presented in Chapter 5. In this case study the link between Communication and Coordination appears to be the most important with respect to breakdowns, and the links

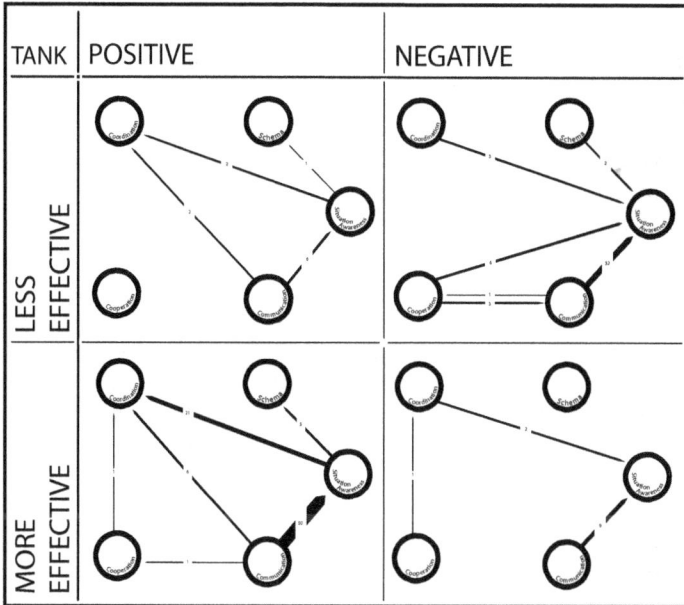

Figure 7.5   Comparison of the less effective and the more effective tank teams' positive and negative links

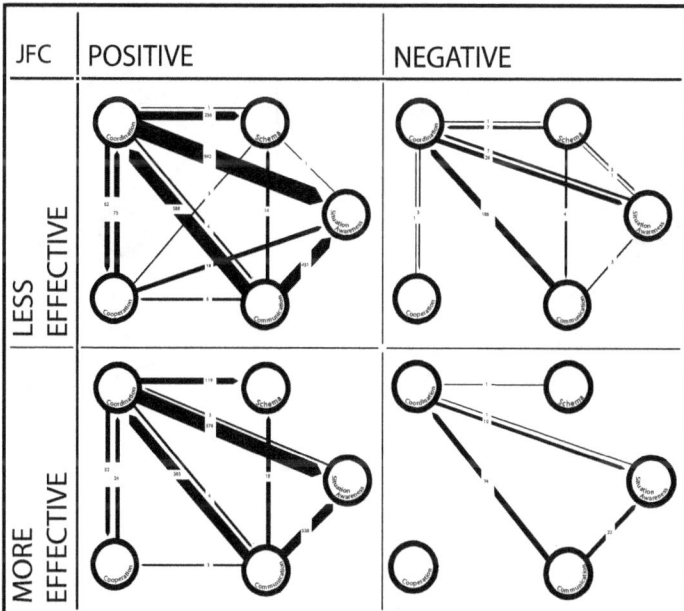

Figure 7.6   Comparison of the less effective and the more effective JFC teams' positive and negative links

between Communication, Coordination and Situation Awareness appear to be the most important with respect to positive links. From this it can be suggested that the links between Situation Awareness, Coordination and Communication represent key causal factors in decision-making within teams. Both teams follow the same pattern of positive links and broken links and both teams have the same hierarchy of frequency for both the broken and positive links. The less effective team, however, has a higher frequency of broken links.

FST Model

Figure 7.7 below presents the data from the Air–Land Integration FST study inputted into the model as presented in Chapter 6. In this case study the link between Communication and Coordination seems to be the most important with respect to breakdowns, and the links between Communication, Coordination and Situation Awareness appear to be the most important with respect to positive links. From this it can be posited that the links between Situation Awareness, Coordination and Communication represent key causal factors in decision-making within teams.

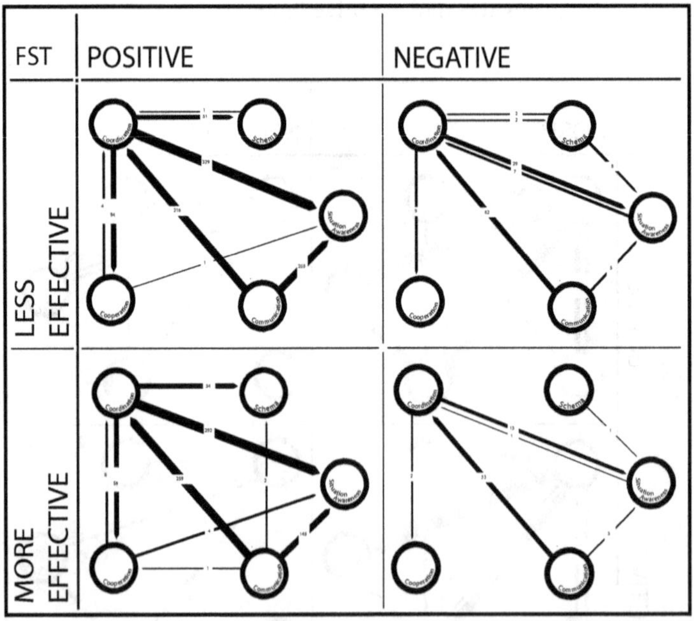

Figure 7.7   Comparison of the less effective and the more effective FST teams' positive and negative links

Comparison of Models

Visual comparison   A surface comparison of the models reinforces the validity of the model derived from the literature. All of the broken links with a frequency of greater than 10 per cent of the total frequency from the data populated models are matched to those links with a value of over 10 per cent of the total frequency represented in the literature populated model. This means that the model derived from the literature is able to explain each of the broken and positive links from the observational case studies. The only exception to this is the link involving the factor Schemata. This suggests, as observed during analysis, that schemata may not be sufficiently measured or represented by the EAST methodology. The debate surrounding the measurement of schemata is complex and ongoing (Langan-Fox, Code and Langfield Smith 2001).

The breakdown in the link between Communication and Situation Awareness is the most prominent in intra-team study. This link is mediated by Coordination for inter-team missions; with Coordination to SA, and Communication to Coordination being the most prominent link breakdowns. The coordination required between teams increases in importance in inter-team case studies, as additional coordination is needed to relay Situation Awareness across multiple teams working together (Fiore, Salas et al. 2003). This notion has been emphasised by a number of other researchers such as Paris, Salas and Cannon-Bowers (2000); Urban, Weaver, Bowers and Rhodenizer (1996); Boiney (2007); and Urban, Weaver et al. (1996). Espinosa, Lerch and Kraut (2002) provide a detailed review of coordination, particularly focusing on the impacts of distribution on coordination. They conclude that distribution makes coordination activity more difficult, and impacts on the manner in which teams coordinate. Espinosa et al. (2002) found that small collocated teams had very different requirements from larger, diverse teams.

Numerical comparison   Exploration of the frequency of links within the models reveals further differences. Initial examination of the total frequency of links within the models illustrates a large difference between the tank, JFC and FST case studies. This is because of the size of the system explored within each case study: the tank study explores a small internal tank crew, whereas the JFC explores a large dual-force organisation. Figure 7.8 represents the frequency of total links within the models, both for positive links and negative links. Figure 7.9 represents the same information portrayed as a percentage of the total number of links for each case study, allowing for a normalised exploration of links across the three case studies.

The figures illustrate that across all case studies there is a greater frequency of negative links within the less effective model when compared with the more effective model. In addition, the figures reveal that there is not always a greater number of positive links in the more effective scenario. In the JFC case study,

Teams	Tank Less Effective	Tank More Effective	JFC Less Effective	JFC More Effective	FST Less Effective	FST More Effective
▫ Positive	11	84	2399	1521	901	821
◼ Negative	47	12	243	108	124	73

Teams

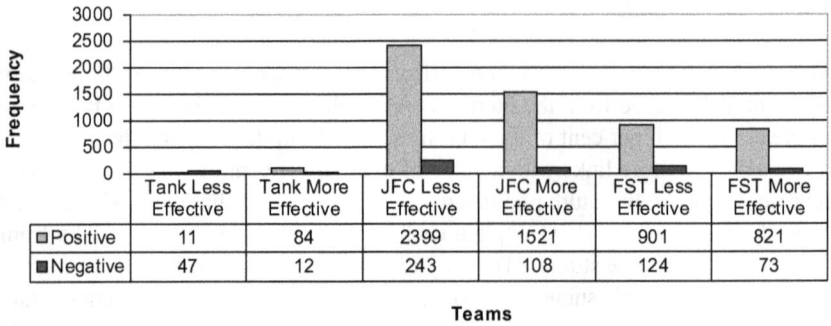

Figure 7.8   Frequency of positive and negative links across all case studies

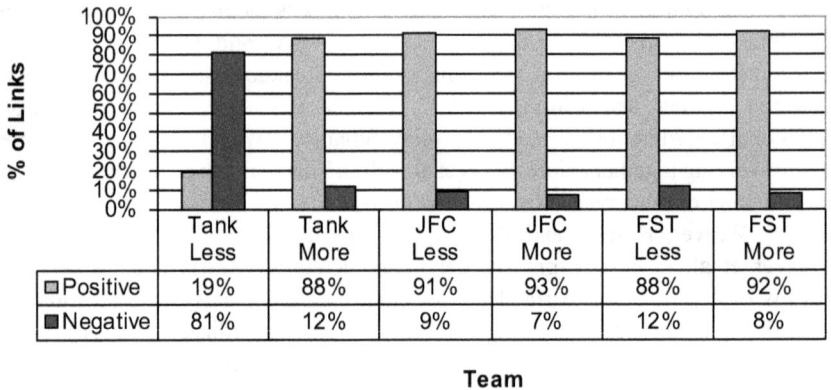

Team	Tank Less	Tank More	JFC Less	JFC More	FST Less	FST More
▫ Positive	19%	88%	91%	93%	88%	92%
◼ Negative	81%	12%	9%	7%	12%	8%

Team

Figure 7.9   Percentage of positive versus negative links for each case study

the less effective team model contains a greater frequency of total positive links compared to the more effective team. From this it can be argued that the presence of negative links is more predictive of fratricide incidents than the presence of positive links. An exploration of the difference in the presence of the links within the models provides support for the influence of negative links over positive links.

Figure 7.10 illustrates the presence of different positive and negative links within the models, that is, how many of the links between the factors are positive, and how many of the links are negative. The less effective models consistently have a higher presence of links which are negative and a greater range of negative links when compared with the more effective models. In line with the frequency of links, the more effective models do not always have a greater presence of positive links. From this it can be put forward that the presence of negative links within a model is more representative of the occurrence of fratricide than the presence of positive links.

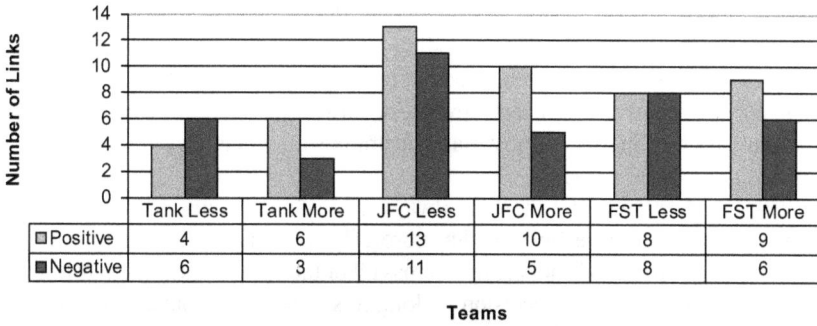

	Tank Less	Tank More	JFC Less	JFC More	FST Less	FST More
Positive	4	6	13	10	8	9
Negative	6	3	11	5	8	6

Teams

Figure 7.10 Presence of positive and negative links across all case studies

In order to explore the impact of the presence of negative links on the occurrence of fratricide, the ratio between positive and negative link presence for each case study are presented below in Figure 7.11, and in Figure 7.12, in a different format.

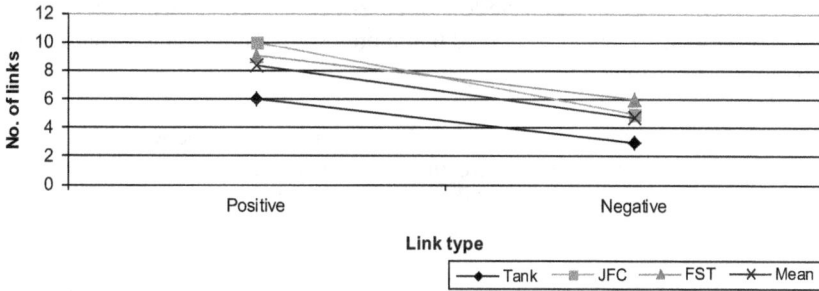

Figure 7.11 Presence of links in the more effective team

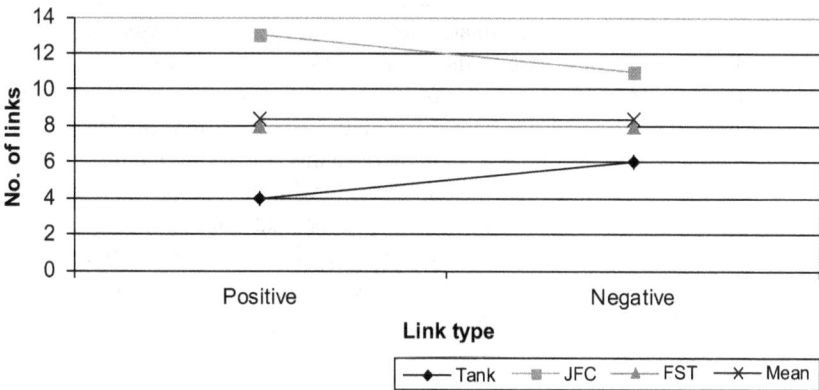

Figure 7.12 Presence of links in less effective team

With the more effective teams, in Figure 7.11, the line on the chart always slopes down from positive to negative, revealing a greater ratio of presence of positive to negative links. With the less effective teams, in Figure 7.12, the lines on the chart do not follow a consistent upward or downward pattern. When the mean is calculated the line sits almost flat, illustrating a similar level of presence of positive to negative links.

The results reveal that there is always a greater *presence* of positive links compared with negative links in the more effective teams. But within the less effective team this ratio is not present across all of the case studies. From this it can be argued that for effective decision-making it is important to have a high ratio of positive to negative links. This again provides support for the influence of negative links over positive links in fratricide incidents.

Summary   Across all of the case studies, the less effective models contain a higher number of broken links, with respect both to type and frequency. That is, the less effective teams have a greater range of broken links, more of the links between its factors are broken, and each of these links has a higher frequency of occurrence when compared with the presence of positive links. The presence of positive links appears to have less of an effect than the presence of negative links, from which it could be argued that more positive links cannot compensate or counteract negative links.

Statistical comparison   In order to explore the interactions between the factors further, graph theory metrics were utilised to compare the populated models statistically.

Sociometric status
Sociometric status is a measure of the level of contribution made by each node to the network (Houghton, Baber et al. 2006). The initial model drawn from the literature presented Communication as the factor with the highest level of sociometric status and as the factor contributing most to determining whether or not an incident of fratricide would occur. An examination of sociometric status across each of the models derived from these case studies is presented in Figure 7.13.

The analysis revealed that in the tank crew case study, Situation Awareness was the node with the highest sociometric status across all four models. Within the FST the node with the highest sociometric status was Coordination across all models, and the same was found within the JFC with the exception of the more effective negative links in which Communication had the highest sociometric status. These results are interesting for a number of reasons. Firstly, within each case study the same node has the highest sociometric status for both the less effective and the more effective teams, and for both positive and negative links (with one exception). From this it can be posited that Coordination and Situation Awareness have such a large impact on the scenarios that they will consistently be the most influential factors.

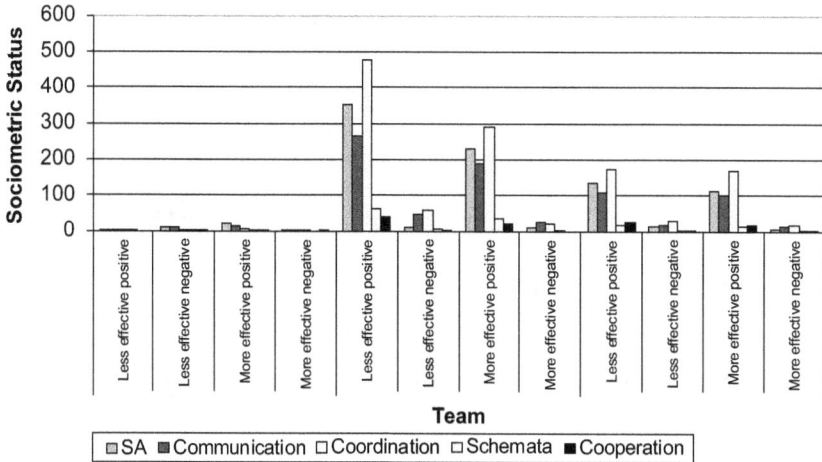

Figure 7.13 Sociometric status

It is also interesting to note that, within the JFC and FST, Coordination is the most significant factor with respect to its contribution to the model. This supports the now weighty claim that Coordination becomes hugely influential within teams of teams. This serves to emphasise the earlier finding that Coordination provides an important mediating role between Communication and Situation Awareness within the multi-team studies of the JFC and FST. The importance of this mediating role is highlighted here by its high sociometric status – it receives an even higher sociometric status than Situation Awareness. From this it can be hypothesised that in small teams Situation Awareness and its associated links are the most influential, whereas within multi-team scenarios Coordination and its associated links become the most influential due to the requirements to coordinate with one another.

The sociometric status drawn from the models also reveals interesting findings regarding Schemata and Cooperation: these two factors held the two lowest sociometric status values across all models in all case studies, the only exception being the negative links in the less effective team in the tank scenario. From this it can be suggested that these two factors and their associated links are the least influential in the occurrence of fratricide incidents. A comparison of the average sociometric status of the models reveals additional differences, as presented in Figure 7.14.

In the tank crew study the more effective positive model had the highest level of mean sociometric status, followed by the less effective negative model, then the more effective negative model followed by the less effective positive model. From this it can be concluded that the most active model was the one with more effective positive links, whereas the least active model was the one with less effective negative links.

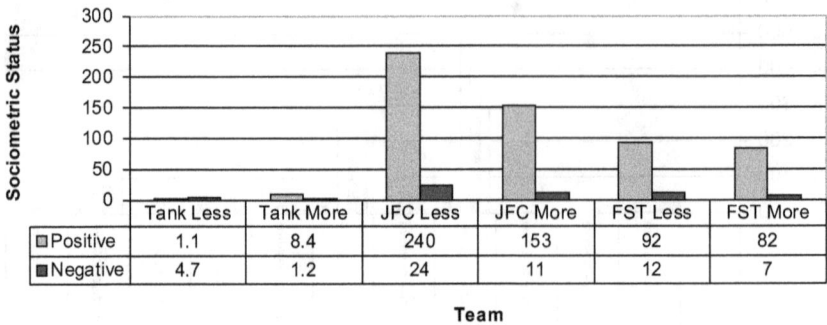

	Tank Less	Tank More	JFC Less	JFC More	FST Less	FST More
☐ Positive	1.1	8.4	240	153	92	82
■ Negative	4.7	1.2	24	11	12	7

Team

Figure 7.14   Average sociometric status

Within the JFC and FST case studies the order of sociometric status values began with the less effective positive model, followed by the more effective positive model, then the less effective negative model and finally the more effective negative model. From this it can be concluded that the sociometric status of negative model outweighs the importance of the sociometric status of the positive model in both of these case studies. Whether or not an incident of fratricide occurs is correlated to the sociometric status of the negative models and not the positive models.

Density
The level of density within a network represents how well connected the network is:

> [T]he overall level of interconnectivity within the network ... (Walker, Stanton and Salmon 2011: 12)

With relation to the F3 model, a higher level of density reflects a greater number of links between the factors. Higher levels of density refer to networks where there is a greater level of direct relational pathways (Walker, Stanton, Salmon and Jenkins 2009). The density values again portray the importance of the negative links in the models. Despite having the same or an even higher level of density within its positive model, the less effective negative model always had a higher value than the more effective model. From this it can be inferred that the model of negative links, or breakdowns, is the model most indicative of fratricide.

The results of the density analysis shown in Figure 7.15 reveal that a well connected negative model of factors undermines a well connected positive model of factors. As is illustrated throughout the analysis, the impact of negative links appears to be disproportionately large compared with the impact of the positive links between factors in the model.

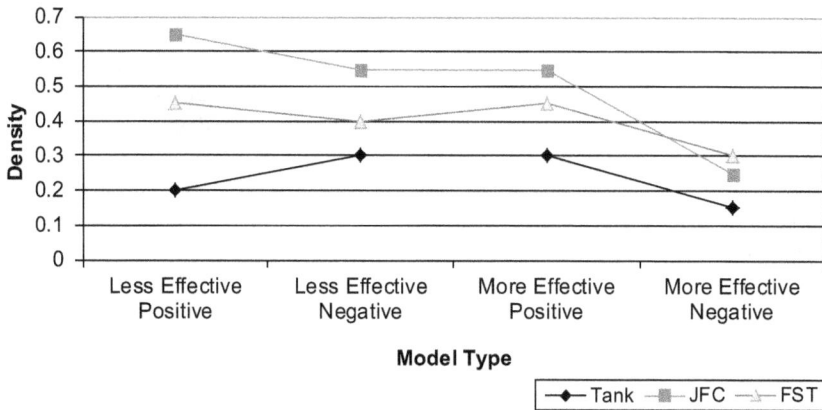

Figure 7.15 Density values across case studies

Summary

The surface, numerical, and statistical comparisons all support the importance of the link between Communication and Situation Awareness, illustrating the mediatory importance of Coordination on this link within large teams of teams. In addition to this, the exploration of the models through metrics has emphasised that the influence of the negative links on whether an incident of fratricide will occur is greater than the influence of positive links. That is, negative links provide a more accurate prediction of a fratricide incident that positive links. The metrics also reinforce the impact of Coordination as a mediatory node within the JFC and FST studies where its influence was higher even than Situation Awareness.

Conclusion

The aim of this chapter was to investigate the similarities and differences between the results of the case studies undertaken within this research. The results of both the coding-based modelling and the EAST analysis reveal a high level of similarity between the two case studies exploring larger teams of teams (JFC and FST) compared to the smaller (within-team) tank crew case study.

In particular, the small within-team case study revealed the benefits of high levels of mutuality with respect to tasks, communication and information. However, in the larger team of teams case studies a more concise approach to tasks, communication and information were seen to be beneficial and a higher emphasis was placed on the coordinated actions required. The results indicate that within small, collocated teams there is a high level of transactive SA, with high levels of communication and information transfer enabling a more accurate situation awareness due to the tight coupling between team members' tasks and environment. Within larger, distributed teams of teams a correlation between effective performance and the principles of DSA was also found, revealing the benefits of efficient information transfer and communication in teams with diverse roles and goals, and the additional coordination demands associated with recognising the information needs of other team members in order to allow efficient communication and information transfer. This research has shown that for small tank crews there is a high level of transactive SA – in that each member of the crew benefited from sharing information they all needed. This position is in line with DSA, not shared SA, in that the individual roles of the tank crew – gunner, driver, and commander – meant that they utilised this information in different manners, and their individual schemata meant that they interpreted the information in different ways (Stanton, Stewart et al. 2006, Stanton, Salmon, Walker and Jenkins 2009a, 2009b, Salmon, Stanton, Walker and Jenkins 2009). It is apparent that there was a tight task coupling within the tank crew, which, along with the geographical collocation meant that a high level of communication and information transfer was beneficial to the crew members.

The research also suggests that the occurrence of fratricide is most affected by the negative actions occurring within the system, regardless of the positive actions taken. That is, the negative actions have a disproportionately detrimental impact on the system. The graph theory metric of density revealed that positive links had less impact on the occurrence of fratricide than negative links. From this it can be proposed that metrics such as density may have a predictive use in the exploration of incidents of fratricide, although further research is needed to explore this link.

The results provide interesting insights into the differences between small intra-teams and larger inter-teams, and between more effective and less effective teams. The analysis highlights the ability of the EAST method to explore the processes used by teams in military decision-making environments. The method provides metrics to measure the factors associated with team performance and could therefore be utilised to compare teams, but also to evaluate the impact of training interventions on team performance. In addition to this, the F3 model provided a framework in understanding both more effective and less effective team performance. This model could be used in the design of training interventions, as well as in developing guidance and decision aids. Extensions to the model and further research could illustrate the utility of the model to explore team decision-making within the military environment and to make predictions about outcomes.

Chapter 8
Conclusions and Recommendations

Introduction

This research set out to explore the problem of fratricide. Previous research into the area emphasised the complexity and multi-causal nature of the problem, illustrating the possibility of exploring fratricide from a systems perspective, incorporating the associated Human Factors issues (Kogler 2003, Jamieson and Wang 2007, Dean and Handley 2006, Gadsen and Outteridge 2006, US Congress 1993, Masys 2006, Wilson et al. 2007, Greitzer and Andrews 2009). The research presented within this book has developed a model to explore fratricide from this perspective, continuing an emphasis on decision-making and teamwork factors associated with fratricide incidents (Wilson et al. 2007, Zobarich, Lamoureux and Bruyn-Martin 2007, Greitzer and Andrews 2009). A systems-based methodology was utilised to populate this model with empirical data, with the objective that such modelling could provide a greater understanding of the processes involved in such incidents.

Summary of Findings

This research has explored the current literature surrounding fratricide and identified a need for a model that explores the problem from a systemic approach, recognising that the causality associated with fratricide is diverse, non-linear and evolutionary. Through the synthesis of the fratricide literature with wider themes in safety and teamwork, a new model was derived to focus examination of fratricide incidents. In order to populate the model with data from incidents of fratricide, a method was developed incorporating Event Analysis of Systemic Teamwork (EAST) and relations coding to evaluate differences between effective performance and incidents of fratricide. After initial validation, the method was applied to three case studies of fratricide drawn from military training institutions, populating the model with empirical evidence.

This research has shown that the performance of more effective teams is correlated with more frequent communications within teams and less frequent communications between teams. These links are characterised by clear communications incorporating efficient information transfer to enable the development of accurate SA. In order to ensure these prominent links occur, additional links are required, including cooperative and coordinative links both within and between teams. The efficiency of these links is further enhanced by the team members' schemata, which enable them to develop accurate expectations of the scenario unfolding around them and appropriate information identification, interpretation and integration. These results provide empirical evidence for the theoretical assertions presented within the fratricide and teamwork literature.

This research highlights the need to ensure that small teams have a strong cohesive environment where they are motivated to, and do, share a high degree of information with one another through frequent communications. They must hold accurate, but efficient, schemata of the situation in order to ensure that information is correctly interpreted and integrated. Within larger teams, it is important to ensure that team members are only provided with appropriate information to reduce cognitive workload and to ensure that relevant information is identified and attended to. Together, these aspects will increase the probability that the system as a whole develops accurate SA and is therefore informed to make an effective decision, regardless of task type, environment and so forth.

The research has illustrated the importance of negative, or broken, links between the F3 model factors. Across the case studies explored, breakdowns, or negative links, consistently had a greater frequency in less effective teams, whereas at points the less effective team also had a greater level of positive links. This suggests that the impact of negative links is disproportionate (in line with systems thinking) and that greater emphasis should be placed on avoiding negative links than on increasing positive links.

A model and a series of metrics to explore the Human Factors issues behind incidents of fratricide, and to make predictions about the future occurrence of

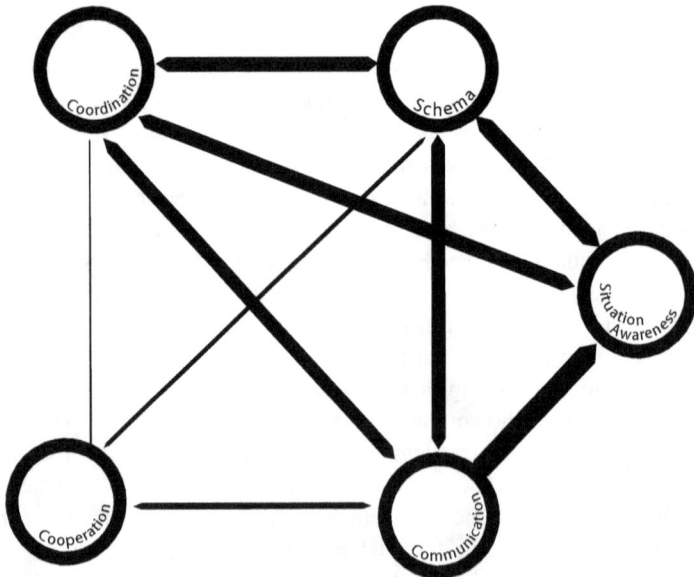

Figure 8.1   F3 model

fratricide, has been developed. The F3 model of fratricide, constructed and provided with initial validation in this research, is presented in Figure 8.1.

The model proposes that incidents of fratricide can be explained by these five factors and their complex relationships with one another. In addition to the factors themselves, the model argues that incidents of fratricide occur due to an inappropriate ratio between positive and negative links between the five factors of the model. Perfect team performance in which no negative links are present cannot be expected, but this research has illustrated a ratio that appears indicative of an increased propensity towards incidents of fratricide.

Main Findings and Relationship with Literature

The Utility of the Systems Perspective

Current research into fratricide has identified the need to explore the interactions between causal factors (Kogler 2003, Jamieson and Wang 2007, Dean and Handley 2006, Gadsen and Outteridge 2006, Gadsen et al. 2008, US Congress 1993, Masys 2006, Wilson et al. 2007). Yet within the literature reviewed there was no model capable of illustrating, or method capable of exploring, these complex interactions. This research has attempted to provide both a model, and a method to enable the exploration of such interactions, continuing the emphasis on exploring fratricide from a systems perspective. The utility of the systems perspective has been proven in numerous domains such as Situation Awareness (Stanton, Stewart et al. 2006, Salmon, Stanton, Walker, Baber, Jenkins, McMaster and Young 2008); Situation Awareness and anaesthesia (Fioratou et al. 2010); Command and Control interface design (Jenkins, Stanton et al. 2008); safety and accidents (Woods et al. 1994, Hollnagel 2004) and human error identification (Stanton and Baber 1996) supporting the research presented here.

This research has provided evidence that empirically supports the theoretical assertions of previous fratricide research; and it is supported by a vast array of literature from wider academia. Zobarich et al. (2007) and Wilson et al. (2007) present initial research into the utility of exploring the phenomenon of fratricide by focusing on team processes, an idea that was continued and emphasised in this research and empirically validated through the case studies undertaken. Perhaps the most important finding of this research is that greater levels of what are generally deemed to be essential team processes are not always beneficial.

Additional Communications Are Not Always Beneficial

Efficient communication is required in order to ensure that members are able to identify, interpret and integrate information appropriately. Within the fratricide domain, research by Hart (2004), Famewo, Matthews and Lamoureux (2007), Bolstad, Endsley and Cuevas (2009), Barnet (2009), Hinsz and Wallace (2009),

Famewo et al. (2007) and Wilson et al. (2007) highlights the risks attached to overloading people with excess communication and information transfer whereby they are unable to decipher relevant and appropriate information within shoot, no-shoot situations. Within the larger team studies – the JFC and FST case studies – a high level of information transfer made it difficult for team members to identify and focus on pertinent information, a problem emphasised by multiple researchers, including Moore et al. (2003) and Hourizi and Johnson (2003) and within the fratricide domain (Bolstad, Endsley and Cuevas 2009, Barnet 2009, Hinsz and Wallace 2009).

Communication was explored in multiple ways within this research. The CUD analysis illustrated the manner in which communication occurred; the SNA illustrated the communication structure utilised; open coding explored the content of communication; and the model-based coding represented the impact of communication on the other causal factors of the F3 model. Analysis of the frequency of communication revealed that between teams a greater level of communication was correlated with less effective performance.

The emphasis on efficient communication and the detrimental impact of excessive communication is emphasised by numerous researchers within the wider error, decision-making and teamwork domains (Moffat 2003a, Mauel 2009, Flin, Slaven and Stuart 1996, Urban et al. 1995, Gorman, Cooke and Winner 2006, Dismukes, Loukopoulos and Jobe 2001, Hutchins, Hocevar and Kemple 1999, Fiore, Salas et al. 2003, Stanton, Stewart et al. 2006). Research has also emphasised that within military situations bandwidth limitations further necessitate the need to ensure efficient communication and information transfer (Walker, Stanton, Jenkins, Salmon and Rafferty 2009, 2010). More efficient communication strategies allow the transfer of relevant information, without overloading the individual with inappropriate information, or masking the relevant information that they need to process in order to achieve effective performance.

The notion of finite cognitive resources is also replicated across the wider literature, for example in Reason's notion of 'bounded rationality' (1990). Humans are unable to process large amounts of information and interpret and integrate this information in an appropriate manner (Reason 1990, Woods et al. 1994); consequently, important information can become masked (Stanton, Stewart et al. 2006). This limit is further emphasised within distributed teams where additional cognitive resources are required to coordinate across the team (Fiore, Salas et al. 2003).

With respect to the communication structure, rather than communication acts, this research has highlighted a complex relationship between communication structure and effective performance. It appears that this relationship is heavily dependent on task factors, communication media and so on. Evidence was found to illustrate a correlation between the presence of a more restricted communication network, over an open network, and effective performance. Literature supporting this finding has hypothesised that teams utilising open communication structures may become overloaded with information processing demands above those imposed on teams within more restricted networks (Omodei, Wearing and McLennan 2000).

Open coding revealed that a greater level of guidance and a lower level of non-mission relevant communications were correlated to effective performance. The wider literature provides support for this finding, hypothesising that non-mission relevant communications provide additional cognitive strain and can distract team members (Dismukes, Loukopoulos and Jobe 2001). Researchers such as Shattuck and Woods (2000) and Pigeau and McCann (2000) emphasise the importance of a high level of guidance, proposing that in order to maintain appropriate action, subordinates 'must understand the supervisors' underlying intent with regard to plans and procedures', especially when supervisors and subordinates are distributed (Shattuck and Woods 2000: 280).

The coding-based modelling revealed that the link between Communication and Situation Awareness is the most important link with respect both to positive and to negative relationships between these factors. That is to say, positive links and negative links between these factors both appear to have a relationship, whether or not the team is involved in an incident of fratricide.

Cooperation can be Disruptive

The research revealed that greater levels of cooperation were not always beneficial in team decision-making scenarios illustrated by higher levels of the SNA metric cohesion in the less effective teams. This result can be explained by literature emphasising the negative impacts of small 'cliques' that could be formed as a result of excessive levels of cohesion (Katz et al. 2005). These cliques are said to have a detrimental impact on the team as a whole by their ability to isolate themselves and create divisions in the larger structure (Katz et al. 2005).

Coordination is Beneficial

A higher quality, as opposed to a higher frequency, of coordination was correlated to effective performance across all of the case studies, and was shown to mediate the link between Communication and Situation Awareness within larger, distributed teams. Coordination has been hypothesised to be positively correlated to effective team performance by researchers such as Stanton and Ashleigh (2000), Entin and Serfaty (1999) and Wilson et al. (2007), and this role is said to be heightened within distributed teams (Fiore, Salas et al. 2003).

In addition to the impact of team processes on fratricide, the research presented here has also illustrated the applicability of DSA (Stanton, Stewart et al. 2006) to explaining a number of differences between more effective and less effective decision-making.

Appropriate Schemata are Essential

Appropriate schemata for the current situation are required in order to ensure that information is correctly interpreted. The illustration of the importance of expectations, or schemata, empirically found within this research also provides strength for the theoretical arguments made in earlier fratricide research (Famewo, Matthews and Lamoureux 2007, Dean and Handley 2006, Wilson et al. 2007, Gadsen et al. 2008, Greitzer and Andrews 2008, 2009).

Within this research it can be seen that inappropriate schemata led to the inaccurate interpretation of information and, indeed, of the situation. The importance of accurate information interpretation was illustrated across all three case studies in which the less effective teams failed to interpret information correctly, resulting in an incident of fratricide. The problems associated with information interpretation have been related to incorrect triggering of schemata. Schemata drive our interaction with and interpretation of the world (Neisser 1976) and, as such, are an important construct in decision-making.

A common theme across the case studies was schemata based around engagement rather than identification processes: the preconception was that the contact was enemy before the identification process began. The presence of the engagement schema meant that the identification process was almost completely side-stepped. The literature surrounding schemata and fratricide theorises that people can ignore information that does not fit with their currently activated schema (Famewo, Matthews and Lamoureux 2007, Dean and Handley 2006, Wilson et al. 2007, Gadsen et al. 2008, Greitzer and Andrews 2009).

The results of the research also suggest that more effective teams are more capable at identifying appropriate information; the information elements present in these teams contains higher levels of relevance than those in the less effective teams.

Tightly coupled with information interpretation is the manner in which information is integrated. The exploration of information integration can reveal insights into mental representations (Walker, Stanton, Salmon, Jenkins, Rafferty and Ladva 2010). It was demonstrated in this research that within the tank crew study effective performance was correlated with a greater level of information integration. Within the JFC and FST case studies the information in the more effective teams was more distributed, illustrating a more distributed awareness of the situation.

Distributed Situation Awareness

The theory of Distributed Situation Awareness (DSA) provides an appropriate explanation of the results presented within this research. The findings related to Situation Awareness and Schemata, and to Communication can be explained in terms of the DSA theory. Stanton, Stewart et al. (2006) argue that, from the DSA perspective, additional communications can be detrimental to the development of

Situation Awareness in three main ways: adding additional, irrelevant information and thus cognitive demands; delaying relevant information transfers; and shadowing the emphasis of relevant information. This is in line with the research presented within this book, which found that efficient communication of key information elements was correlated with effective performance.

DSA contends that agents within a team have different roles and schemata and, therefore, utilise and interpret information in different ways. Despite such differences, not all SA, or information, required by team members is diverse. Within a team there may be an exchange between 'transactive SA', that is, information elements that are required by multiple team members. Within a small team, where roles and tasks are tightly coupled, there is a higher number of transactive information elements; as a result, a high level of communication is required in order to ensure that each team member receives the correct information. Within larger teams, where roles and tasks are more loosely coupled, there are lower levels of transactive information elements; therefore, a more efficient level of communication is required in order to ensure that only appropriate information is passed and team members do not become overloaded and relevant information does not become masked. The F3 model was able to illustrate the manner in which the factors interact and the theory of DSA provides an explanation as to why the interaction has occurred in this particular manner in the case studies explored in this book.

Core Issues

The aim of this research was to explore the underlying Human Factors associated with incidents of fratricide. In order to do this, the performances of teams involved in incidents of fratricide were compared to those of teams that performed effectively. The analysis revealed key differences between more effective and less effective (as defined by the occurrence of fratricide) teams. The findings derived from the analysis of the three case studies were used to identify the various issues that are likely to hinder shoot, no-shoot decision-making within teams. From the research presented within this book a number of core principles have been identified to aid in the reduction of fratricide incidents:

- Non-mission relevant communications should be minimised.
- The frequency of communications should be increased within small collocated teams.
- A hierarchical communication structure should be utilised within small collocated teams.
- The frequency of communications should be reduced within large distributed teams of teams.
- The number of task steps taken to complete goals or sub-goals should be reduced to an efficient number.

- Cooperation, both within and between teams, should be encouraged.
- Highly integrated SA, with shorter paths between information elements, should be fostered in small, collocated teams.
- Distributed SA, characterised by longer paths between information elements, should be fostered in larger distributed teams of teams.
- The number of inappropriate task steps taken should be reduced.
- The quality of coordinated activities should be improved.
- It should be ensured that appropriate expectations are developed.
- Relevant information should be made more salient.

These principles have been derived from the research presented in Chapter 4, Chapter 5 and Chapter 6. Their applicability to teams is dependent upon the make up of the team: whether the team is a small, collocated team (the three-man tank team in Chapter 4); or a larger, distributed team of teams (the JFC and FST in Chapter 5 and Chapter 6). Table 8.1 below presents the principles along with an illustration of the team type they are applicable to (shaded cells represent principles relevant to that team type).

It could be argued that the conclusions and recommendations are not unique to this book, many are emphasised across the wider team and error domains and within the fratricide literature itself (Doton 1996, US Congress Office of Technology Assessment 1993, Hart 2004, Wilson et al. 2007, Wilson, Salas and Andrews 2009, Gadsen et al. 2008, Greitzer and Andrews 2008, Famewo, Matthews and Lamoureux 2007, Famewo et al. 2007, Dean and Handley 2006, Masys 2006). Within this research, however, the conclusions are drawn from an in-

Table 8.1   Principles for the reduction of fratricide in teams

	Small collocated team	Large distributed team of teams
Reduce non-mission relevant comms		
Increase frequency of comms		
Utilise a hierarchical communication structure		
Reduce frequency of comms		
Reduce number of tasks per goal		
Encourage cooperation		
Foster integrated SA		
Foster distributed SA		
Reduce the number of inappropriate tasks		
Improve quality of coordination		
Ensure development of appropriate expectations		
Make relevant information more salient		

depth empirical evaluation of the problem of fratricide. The conclusions presented here illustrate the applicability of a number of generic Human Factors issues to the problem of fratricide and depict the specific way in which they manifest themselves within this domain. Such empirical analysis provides the conclusions and recommendations presented here with a higher degree of validity.

Utility

In the Introduction to this book and in Chapter 1, the current emphasis on technological solutions to fratricide is highlighted. A number of researchers have argued against such a technological focus, stating that solutions must be based on a firm foundation of knowledge surrounding the decision-making process in incidents of fratricide (Jamieson and Wang 2007, Wilson et al. 2007, Parasuraman and Riley 1997, Dzindolet et al. 2001, Doton 1996). Without such knowledge, technological solutions may be inappropriate (Hart 2004), fundamentally alter the process (Masys 2006), and may be inappropriately utilised, whether this be misuse, abuse or disuse (Dzindolet et al. 2009). Fernall (1997) highlights the need to understand the way in which military personnel make decisions before attempting to develop decision support aids, illustrating numerous examples of decision support failures that arose due to an inaccurate understanding of the human involved. Fernall further argues that research must explore the 'factors affecting decision making; relationships between decision making and decision support; and effective methods for generating support requirements' (1997: 215).

This research aimed to provide such a foundation of knowledge. Exploring and modelling decision-making in complex teams within naturalistic environments provides valuable information that can be used not only to create effective and appropriate tools and decision aids for soldiers but also to devise appropriate rules, procedures, tactics, training, and so on, to place them in the best possible position to make effective shoot, no-shoot decisions.

Conclusions

From the summation of this research, a theoretical model of the core factors affecting fratricide, and the manner in which these factors interact, has been developed and validated. This model has been validated through its application to three military case studies and its ability to explore both fratricide incidents and effective performance within shoot, no-shoot decisions. Empirical evidence has been derived to illustrate the applicability of the method, model and systems perspective in the exploration of incidents of fratricide. In addition to this, current research into fratricide, emphasising the role of teamwork, expectations, SA and multiple interacting causal factors, has been supported. The combination of these populated models and the EAST analysis has allowed the identification of core issues faced by teams undertaking shoot, no-shoot decisions.

Future Research

It is hoped that future work will continue to build upon the model and apply the method to multiple diverse scenarios, including the following.

Naval Operations

This research was unable to explore an in-depth analysis of an incident of fratricide within the Royal Navy. A number of observations have been undertaken by the authors exploring pre-deployment training in shoot, no-shoot decisions in the naval domain. These observations enabled a high-level understanding of the processes involved through which numerous crossovers with Army and RAF decision-making were highlighted. It is the authors' opinion that application of the F3 model and the EAST method to the naval domain would provide further support for both the model and the methodology, as well as interesting insights into the processes involved in the Royal Navy.

Tri-force Operations

The research presented in this book has explored fratricide scenarios within Joint Force environments between the British Army and the RAF. The model could also be applied to tri-force scenarios incorporating all three of the British armed forces. It is posited that the applicability of the F3 model and the EAST method would be further illustrated within such a domain. Application to a tri-service domain would enable an evaluation of the complexity explored by the method and the model, and an assessment of whether the appropriate level of complexity is provided for tri-force operations.

Multinational Operations

In addition to exploring tri-force operations, it is hoped that the model can be applied to multinational operations. The military operations that Britain is currently involved in comprise international forces made up of many different nations. Within these contexts, research has suggested that fratricide rates could increase due to differences in equipment, tactics, procedures and so on between the forces, in addition to an even greater coordination demand when synchronising such a great number of forces. In line with research into tri-force operations, application of the method and model to multinational operations will ensure that both contain the appropriate level of complexity required for this level of analysis.

Applicability to Wider Decision-making Tasks in the Military

The focus of this book was on the decision-making cycle involved in incidents of fratricide. During the research, a decision-making scenario was explored that did not involve engagement decisions: a squad of helicopters was observed undertaking a mission in which the decision-making broke down, resulting in the helicopters following an incorrect and potentially dangerous route (Stanton, Rafferty et al. 2010). The core factors of the model were shown to be present in this scenario and allowed for an explanation of the decision-making breakdown. Further research is needed to explore the applicability of the F3 model to non-engagement based decision-making processes.

Diverse Domains

The research presented within this book is based around the military domain; however, the focus of the research has been on the team processes involved in decision-making, rather than on the military context. Due to this, the F3 model may have applicability in numerous domains involving difficult decision-making tasks. The results of this research could be transferred to other safety-critical domains already highlighted in Human Factors research, such as offshore oil platforms (Flin, Slaven and Stewart 1996), police shooting decisions (Jenkins, Salmon et al. 2011, Flin, Pender, Wujec, Grant and Stewart 2006) and aviation accidents (Griffin, Young and Stanton 2010).

Team Evolution

These results are drawn from the initial stages of an intense pre-deployment training series and all conclusions must be taken with this in mind. Future work aims to compare performance at the beginning of training and post-completion of training to identify whether these issues are still present, or are negated through the training.

Research has shown that teams evolve over time (Hollingshead et al. 2005) and this research explores only a snapshot in time for each team. It would be interesting to explore the manner in which team evolution impacts on the F3 model. Future research could explore teams from initial composition across the training regime to identify the way in which the teams' evolution affects the F3 model, the decision-making processes, and to explore the utility of specific training interventions.

Extended Measurement of Cooperation

During this research, exploration of the concept of cooperation and its interaction with other factors was challenging. Future research should aim to solidify an appropriate methodology to explore the attitudinal aspects of teamwork and its impact on team processes. Previous research into this area has emphasised the utility of tools such as the Cockpit Management Attitudes Questionnaire (CMAQ) to measure team attitudes in teams in complex domains such as surgery (Flin, Fletcher et al. 2003), offshore oil drilling (Crichton 2005) and aviation (Helmreich 1984). Due to access restrictions, the application of such questionnaires was prohibited within this research but could provide interesting additional insights in future research.

Extended Schemata Measurement

The research presented within this book highlights the need for a reliable, valid and non-subjective measure of schemata. The use of Leximancer made it possible to extract core thematic groupings, providing an initial exploration of the integration and interpretation of information from which schemata can be determined. Subjective analysis of the data also enabled insights into the relevant schemata utilised within the scenario. However, as Langan-Fox and colleagues have argued, a faultless measure of schemata has so far 'eluded researchers' (Langan-Fox et al. 2001: 99). Neisser argues that 'schemata are locked inscrutably within his skull where we cannot see them' (1976: 186). It is hoped that this book will stimulate other researchers to explore the problem and consider the development of new measurement techniques.

Predictive Utility of Model

The current research has focused on the retrospective exploration of fratricide incidents. It would be interesting to utilise the F3 model and the findings derived from the case studies in order to predict incidents of fratricide. Specifically, the ratio between positive and negative links for the social network density diagnostic appears to correlate with the occurrence of an incident of fratricide, as well as the centrality metric for the information networks. The predictive utility of the method and model could be used to develop a series of specific training requirements for teams during pre-deployment training. In addition to this the metrics could be used to analyse numerous teams in order to attempt to predict the occurrence of incidents of fratricide.

Closing Remarks

This research began in 2007 with an initiation to military decision-making on a three-week observation of a military trial in Germany. Meeting soldiers in the field, experiencing the problems they face on a day-to-day basis and understanding the commitment with which they risk their lives emphasised the importance of the issue of fratricide. This book set out to explore the area by preparing to conduct a number of experimental trials; however, the research took a more naturalistic approach rooted in grounded theory as a result of standing alongside troops in a number of training scenarios and watching fratricide events unfold. The observations undertaken substantiated the meaning of the expression 'fog of war' (Clausewitz 1873). Even within simulated environments, the complexity, pressure and multiple stimuli present were overwhelming to observe, let alone process. In such situations making a decision is never simple.

This research has presented a theoretical framework and empirical methodology that, it is hoped, will enable further exploration of fratricide incidents. The devastation of such incidents necessitates further research within this area.

References and Bibliography

Adelman, L. (1991). Experiments, quasi-experiments, and case studies: A review of empirical methods for evaluating decision support systems. *IEEE Transactions on Systems, Man and Cybernetics*, 21(2), 293–301.

Alberts, D.S. and Hayes, R.E. (2003). *Power to the Edge: Command and Control in the Information Age*. Washington, DC: CCRP Press.

Alexander, R., Hall-May, M. and Kelly, T. (2004). Characterisation of Systems of Systems Failures. Proceedings of the 22nd International System Safety Conference. Available at: http://www-users.cs.york.ac.uk/~tpk/issc04c.pdf

Almeida, I.M. and Johnson, C.W. (2005). Extending the Borders of Accident Investigation: Applying Novel Analysis Techniques to the Loss of the Brazilian Space Programme's Launch Vehicle VLS-1 V03. [Online]. Available at: http://www.dcs.gla.ac.uk/~johnson/papers/Ildeberto_and_Chris.PDF

Andrews, D.H. (2009). Preface, in *Human Factors in Combat Identification*, edited by D.H. Andrews, P.H. Robert and M.B. Wolf. Aldershot: Ashgate.

Annett, J. (2005). Hierarchical task analysis, in *Handbook of Human Factors and Ergonomics Methods*, edited by N.A. Stanton et al. London: CRC Press.

Annett, J. and Stanton, N.A. (2000). Editorial: Teamwork – A problem for ergonomics. *Ergonomics*, 43(8), 1045–51.

Annett, J., Cunningham, D. and Mathias-Jones, P. (2000). A method for measuring team skills. *Ergonomics*, 43(8), 1076–94.

Antoni, C.H. (2006). Teamwork and sociotechnical systems theory, in *International Encyclopaedia of Ergonomics and Human Factors*, edited by W. Karwowski. Bristol, PA: Taylor and Francis.

ARRSE (2010). *Fire Support Team*. [Online]. Available at: http://www.arrse.co.uk/wiki/Fire_Support_Team

Atkinson, S.R. and Moffat, J. (2005). *The Agile Organisation: From Informal Networks to Complex Effects and Agility*. Washington, DC: CCRP Press.

Ayeko, M. (2002). Integrated Safety Investigation Methodology (ISIM) – Investigating for Risk Mitigation, in Workshop on the Investigation and Reporting of Incidents and Accidents (IRIA), Glasgow, 17th–20th July 2002, edited by C.W. Johnson. Available at: http://www.dcs.gla.ac.uk/~johnson/iria2002/IRIA_2002.pdf

Baber, C. and Mellor, B.A. (2001). Modelling multimodal human–computer interaction using critical path analysis. *International Journal of Human Computer Studies*, 54(4), 613–36.

Baber, C. and Stanton, N.A. (1996). Observation as a technique for usability evaluations, in *Usability in Industry*, edited by P.W. Jordan. London: Taylor and Francis.

Baber, C. and Stanton, N.A. (2004). Methodology for DTC-HFI WP1 Field Trials. HFI DTC Technical Report/WP.2.1.

Baber, C., Houghton, R.J., McMaster, R., Salmon, P., Stanton, N.A., Stewart, R.J. and Walker, G. (2004). Field Studies in Emergency Services. HFI DTC Technical Report/WP1.1.1/01.

Baber, C., Walker, G., Stanton, N.A. and Salmon, P. (2004). Report on Initial Trials of WP 1.1 Methodology Conducted at Fire Service Training College. HFI DTC Technical Report/WP 1.1.1/1.1.

Baker, D.P., Day, R. and Salas, E. (2006). Teamwork as an essential component of high reliability organisations. *Health Services Research*, 41(4), 1576–98.

Baker, D.P. and Krokos, K.J. (2007). Development and validation of Aviation Causal Contributors for Error Reporting Systems (ACCERS). *Human Factors*, 49(2), 185–99.

Balakrishnan, J.D. and McDonald, J.A. (2006). Signal Detection Theory, in *International Encyclopaedia of Ergonomics and Human Factors*, edited by W. Karwowski. Bristol, PA: Taylor and Francis.

Banbury, S., Dudfield, H. and Lodge, M. (2002). Development and Preliminary Validation of a Cognitive Model of Commercial Airline Pilot Threat Management Behaviour, in Workshop on the Investigation and Reporting of Incidents and Accidents (IRIA), Glasgow, 17th–20th July 2002, edited by C.W. Johnson. Available at: http://www.dcs.gla.ac.uk/~johnson/iria2002/IRIA_2002.pdf

Barnet, J. (2009). The case for active fratricide avoidance in net-centric C2 systems, in *Human Factors in Combat Identification*, edited by D.H. Andrews, P.H. Robert and M.B. Wolf. Aldershot: Ashgate.

Bar Yam, Y. (1997). *Dynamics of Complex Systems*. Jackson, TN: Perseus.

Bar Yam, Y. (2003). Complexity of Military Conflict: Multiscale Complex Systems Analysis of Littoral Warfare. Report for Chief of Naval Operations Strategic Studies Group. [Online.] Available at: http://necsi.edu/projects/yaneer/SSG_NECSI_3_Litt.pdf

Bartlett, F.C. (1932). *Remembering: A Study in Experimental and Social Psychology*. Cambridge: Cambridge University Press.

Baysari, M.T., McIntosh, A.S. and Wilson, J.R. (2008). Understanding the Human Factors contribution to railway accidents and incidents in Australia. *Accident Analysis and Prevention*, 40, 1750–57.

Beach, L. and Lipshitz, R. (1993). Why classical decision theory is an inappropriate standard for evaluating and aiding most human decision making, in *Decision Making in Action: Models and Methods*, edited by G.A. Klein, J. Orasanu, R. Calderwood and C.E. Zsambok. Norwood, NJ: Ablex.

Benner, L. (1979). Crash theories and the implications for research. *American Association of Automitive Medicine Quarterly Journal*, 1(1). [Online]. Available at: http://members.cox.net/lbjr05/papers/AAAM.html

Benner, L. (1985). Rating accident models and investigation methodologies. *Journal of Safety Research*, 16, 105–26.

Benta, M.I. (2003). Agna 2.1 User's Manual. [Online]. Available at: www. geocities.com/imbenta/agna

Boiney, L.G. (2007). More than Information Overload: Supporting Human Attention Allocation. 12th International Command and Control Research and Technology Symposium, Newport, Rhode Island, USA, June 19th–21st, 2007. Available at: www.dodccrp.org

Bolstad, C.A., Endsley, M.R. and Cuevas, H.M. (2009). Team coordination and shared situation awareness in combat identification, in *Human Factors in Combat Identification*, edited by D.H. Andrews, P.H. Robert and M.B. Wolf. Aldershot: Ashgate.

Bowers, C.A., Urban, J.M. and Morgan Jnr, B.B. (1992). *The Study of Crew Coordination and Performance in Hierarchical Team Decision Making*. University of Central Florida, Team Performance Lab, Technical Report 92–01. Available at: http://www.dtic.mil/cgi-bin/GetTRDoc?Location=U2&doc= GetTRDoc.pdf&AD=ADA259514

Braband, J. and Brehmke, B. (2002). Why because graphs Application of Why-Because Graphs to Railway Near-Misses, in Workshop on the Investigation and Reporting of Incidents and Accidents (IRIA), Glasgow, 17th–20th July 2002, edited by C.W. Johnson. Available at: http://www.dcs.gla.ac.uk/~johnson/ iria2002/IRIA_2002.pdf

Braband, J., Evers, B. and Stefano, E. (2003). Towards a hybrid approach for incident root cause analysis. Siemens Transportation Systems – Rail Automation System Development Integrity. Available at: http://rzv113.rz.tu-bs.de/Bieleschweig/pdfB2/deStefano_Bieleschweig.pdf

Briscoe, G.J. (1990). Mort Based Risk Management. Safety System Development Center, EG&G Idaho Inc. Idaho National Engineering Laboratory.

Bryant, D.J. (2006). Rethinking OODA: Toward a modern cognitive framework of command decision making. *Military Psychology*, 18(3), 183–206.

Bundy, G. (1994). Not so friendly fire: Considerations for reducing the risk of fratricide. Newport, R.I: Naval War College Department of Operations: Final report. Available at: http://www.dtic.mil/cgi-bin/GetTRDoc?AD=ADA28350 1&Location=U2&doc=GetTRDoc.pdf

Burke, S.C. (2005). Team task analysis, in *Handbook of Human Factors and Ergonomics Methods*, edited by N.A. Stanton et al. London: CRC Press.

Burt, R.S. (1982). *Toward a Structured Theory of Action: Network Models of Social Structures, Perception and Action*. New York: Academic Press.

Cahillane, M.A., Morin, C., Whitworth, I.R., Farmilo, A. and Hone, G. (2009). Examining the Perception of Miscommunication in a Coalition Environment, 14th International Command and Control Research and Technology Symposium, Washington DC, USA, June 15th–17th, 2009. Available at: http:// www.dodccrp.org/events/14th_iccrts_2009/presentations/072.pdf

Cannon-Bowers, J.A. and Salas, E. (1998). *Making Decisions Under Stress*. Washington, DC: American Psychological Association.

Cannon-Bowers, J.A., Salas, E. and Converse, S. (1993). Shared mental models, in *Individual and Group Decision Making*, edited by J.N. Castellan. Hillsdale, NJ: Lawrence Erlbaum Associates.

Cannon-Bowers, J.A., Salas, E., Blickensderfer, E. and Bowers, C.A. (1998). The impact of cross-training and workload on team functioning: A replication and extension of initial findings. *Human Factors*, 40, 92–101.

Carayon, P. (2006). Human Factors of complex sociotechnical systems. *Applied Ergonomics*, 37(4), 525–35.

Carstens, D.S. (2006). Error analysis leading technology development. *Theoretical Issues in Ergonomics Science*, 9(6), 479–99.

Center for Army Lessons Learned (CALL) (2005). Dispatches: Lessons learned for soldiers: Fratricide. 11(1). Fort Leavenworth, KA: CAC (United States Army Combined Arms Centre). Available at: http://armyapp.dnd.ca/allc/Downloads/dispatch/Vol_11/Dispatches_%20Vol_11%20No_1English.pdf

Clausewitz, C. Von. (1873). *On War*. London: N. Trübner. Available at: http://www.clausewitz.com/readings/OnWar1873/BK2ch02.html

Converse, S.A., Cannon-Bowers, J.A., and Salas, E. (1991). Team member shared mental models: A theory and some methodological issues. Human Factors Society 35th Annual Meeting, San Francisco, CA, USA, September 2nd–6th, 1991.

Cooke, H.J. and Gorman, N.J. (2006). Assessment of team cognition, in *International Encyclopaedia of Ergonomics and Human Factors*, edited by W. Karwowski. Bristol, PA: Taylor and Francis.

Cooke, N.J., Salas, E., Cannon-Bowers, J.A. and Stout, R.J. (2000). Measuring team knowledge. *Human Factors*, 42(1), 151–73.

Coury, B.G. and Terranova, M. (1991). Collaborative decision making in dynamic systems. Human Factors Society 35th Annual Meeting, San Francisco, CA, USA, September 2nd–6th, 1991.

Crandall, B., Klein, G. and Hoffman, R.R. (2006). *Working Minds: A Practitioner's Guide to Cognitive Task Analysis*. Cambridge, MA: MIT Press.

Crichton, M. (2005). Attitudes to teamwork, leadership, and stress in oil industry drilling teams. *Safety Science*, 43, 679–96.

Crichton, M.T. and Flin, R. (2004). Identifying and training non-technical skills of nuclear emergency response teams. *Annals of Nuclear Energy*, 31, 1317–30.

Crichton, M., Flin, R. and McGeorge, P. (2005). Decision making by on-scene commanders in nuclear emergencies. *Cognition, Technology and Work*, 7, 156–66.

The *Daily Mail* (2008). Nine paratroopers shot by British gunship after being mistaken for Taliban. July 11th, 2008. [Online]. Available at: http://www.dailymail.co.uk/news/article-1034126/Nine-Paras-shot-British-gunship-mistaken-Taliban-friendly-disaster.html

Darke, P., Shanks, G. and Broadbent, M. (1998). Successfully completing case study research: Combining rigour, relevance and pragmatism. *Information Systems Journal*, 8, 273–89.

Dean, D.F. and Handley, A. (2006). *Representing the Human Decision Maker in Combat Identification.* CCRTS: The State of the Art and the State of Practice. Available at: http://www.dtic.mil/cgi-bin/GetTRDoc?AD=ADA463172&Location=U2&doc=GetTRDoc.pdf

Dekker, A.H. (2002). C4ISR Architectures, Social Network Analysis and the FINC Methodology: An Experiment in Military Organisational Structure. Information Technology Division Electronics and Surveillance Research Laboratory DSTO-GD-0313. Available at: http://www.dtic.mil/cgi-bin/GetTR Doc?AD=ADA403027&Location=U2&doc=GetTRDoc.pdf

Dekker, S. (2003). Illusions of explanation: A critical essay on error classification. *The International Journal of Aviation Psychology*, 13(2), 95–106.

Dekker, S. (2005). *Ten Questions about Human Error*. Aldershot: Ashgate.

Dekker, S.W.A. and Waldmann, T. (2006). Predicting design induced Pilot error using HET (Human Error Template) – a new formal Human Error Identification method for flight decks. *The Aeronautical Journal of the Royal Aeronautical Society*, February, 107–15.

Dempsy, P.G. (2006). Accident and incident investigation,in *Handbook of Human Factors and Ergonomics*, edited by G. Salvendy. Third edition. Hoboken, NJ: John Wiley and Sons Inc.

Denscombe, M. (2007). *The Good Research Guide: For Small Scale Social Science Research Projects*. Maidenhead: McGraw Hill International.

Derosier, C., Leclercq, S., Rabardel, P. and Langa, P. (2008). Studying work practices: A key factor in understanding accidents on the level triggered by a balance disturbance. *Ergonomics*, 51(12), 1926–43.

Dismukes, R.K., Loukopoulos, L.D. and Jobe, K.K. (2001). The Challenges of Managing Concurrent and Deferred Tasks. 11th International Symposium on Aviation Psychology, Columbus, Ohio: Ohio State University, edited by R. Jensen. Available at: http://human-factors.arc.nasa.gov/publications/KD_LL_KJ_ISAP01.pdf

Doton, L. (1996). Integrating technology to reduce fratricide. *Acquisition Review Quarterly*, Winter. Available at: http://www.dtic.mil/cgi-bin/GetTRDoc?Location=U2&doc=GetTRDoc.pdf&AD=ADA487939

Drillings, M. and Serfaty, D. (1997). Naturalistic decision making in command and control, in *Naturalistic Decision Making*, edited by C.E. Zsambok and G.A. Klein. Mahwah, NJ, Lawrence Erlbaum Associates Inc.

Driskell, J.E. and Mullen, B. (2005). Social network analysis, in *Handbook of Human Factors and Ergonomics Methods*, edited by N.A. Stanton et al. London: CRC Press.

Driskell, J.E. and Salas, E. (1991). Group decision making under stress. *Journal of Applied Psychology*, 76, 473–8.

Dzindolet, M.T., Pierce, L.G. and Beck, H.P. (2009). An examination of the social, cognitive, and motivational factors that affect automation reliance, in *Human Factors in Combat Identification*, edited by D.H. Andrews, P.H. Robert and M.B. Wolf. Aldershot: Ashgate.

Eisenhardt, K.M. (1989). Building theories from case study research. *The Academy of Management Review*, 14(4), 532–50.

Endsley, M.R. (1995). Towards a theory of Situation Awareness in dynamic systems. *Human Factors*, 37, 32–64.

Entin, E.E. and Serfaty, D. (1999). Adaptive team coordination. *Human Factors*, 41(2), 312–25.

Espinosa, A., Lerch, J. and Kraut, R. (2002). Explicit vs. implicit coordination mechanisms and task dependencies: One size does not fit all. [Online]. Available at: http://www.cs.cmu.edu/~kraut/RKraut.site.files/articles/Espinosa 03-ExplicitVsImplicitCoordination.pdf

Famewo, J.J., Bruyn-Matrin, L.E., Zobarich, R.M., Vilhena, P.G.S. and Lamoureux, T.M. (2007). Combat identification: A summary of the literature, function flow analysis and decision requirements analysis. DRDC Toronto, No. CR-2007–123. [Online]. Available at: http://cradpdf.drdc.gc.ca/PDFS/unc69/p528795.pdf

Famewo, J.J., Matthews, M. and Lamoureux, T.M. (2007). Models of information aggregation pertaining to combat identification: A review of the literature. DRDC Toronto. No. CR-2007–062. [Online]. Available at: http://www.dtic. mil/cgi-bin/GetTRDoc?AD=ADA482371&Location=U2&doc=GetTRDoc.pdf

Fan, X., McNeese, M. and Yen, J. (2010). NDM-Based cognitive agents for supporting decision-making teams. *Human-Computer Interaction*, 25(3), 195–234.

Farrington-Darby, T., Wilson, John R., Norris, B. J. and Clarke, T. (2006). A naturalistic study of railway controllers. *Ergonomics*, 49(12), 1370–94.

Fernall, R. (1997). Military decision support, in *Decision Making Under Stress*, edited by R. Flin, E. Salas, M. Strub and L. Martin. Aldershot: Ashgate.

Ferry, T.S. (1988). *Modern Accident Investigation and Analysis*. New York: John Wiley – IEEE.

Fioratou, E., Flin, R., Glavin, R. and Patey, R. (2010). Beyond monitoring: Distributed Situation Awareness in anaesthesia. *British Journal of Anaesthesia*, 105(1), 83–90.

Fiore, S.M., Rosen, M.A., Salas, E., Letsky, M. and Warner, N. (2008). Macrocognition in Command and Control: Understanding and Assessing Verbal and Non-verbal Communications During Complex Collaborative Problem Solving. 13th International Command and Control Research and Technology Symposium, Bellevue, WA, USA. June 17th–19th, 2008. Available at: www. dodccrp.org

Fiore, S.M., Salas, E., Cuevas, H.M. and Bowers, C.A. (2003). Distributed coordination space: Toward a theory of distributed team process and performance. *Theoretical Issues in Ergonomic Science*, 4(3–4), 340–64.

Fischer, U., McDonnell, L. and Orasanu, J. (2007). Linguistic correlates of team performance: Toward a tool for monitoring team functioning during space missions. *Aviation, Space and Environmental Medicine*, 78(5), Section II.

Fleishman, E.A. and Zaccaro, S.J. (1992). Toward a taxonomy of team performance functions, in *Teams: Their Training and Performance*, edited by R.W. Swezey and E. Salas. Norwood, NJ: Ablex.

Flin, R. (1997). Crew resource management for teams in the offshore oil industry. *Team Performance Management*, 3(2), 121–9.

Flin, R. and Maran, N. (2004). Identifying and training non-technical skills for teams in acute medicine. *Qualitative Safety Health Care*, 13(suppl.1), i80–i84.

Flin, R., Fletcher, G., McGeorge, P., Sutherland, A. and Patey, R. (2003). Anaesthetists' attitudes to teamwork and safety. *Anaesthesia*, 58, 233–42.

Flin, R., Salas, E., Strub, M. and Martin, L. (1997). Introduction, in *Decision Making Under Stress*, edited by R. Flin, E. Salas, M. Strub and L. Martin. Aldershot: Ashgate.

Flin, R., Slaven, G. and Stewart, K. (1996). Emergency decision making in the offshore oil and gas industry. *Human Factors* 38(2), 262–77.

Flin, R., Yule, S., McKenzie, L., Paterson-Brown, S. and Maran, N. (2006). Attitudes to teamwork and safety in the operating theatre. *Surgeon*, 4(3), 145–51.

Flin, R., Pender, Z., Wujec, L, Grant, V. and Stewart, E. (2007) Police officers' assessment of operational situations. *Policing*, 30, 310–23.

Flyvberg, B. (2006). Five misunderstandings about case study research. *Qualitative Inquiry*, 12(2), 219–45.

Freeman, J.T. and Cohen, M.S. (1994). Training Metacognitive Skills for Situation Awareness. Symposium on Command and Control Research and Decision Aids. Monterey, CA, June 1994. Available at: http://www.cog-tech.com/papers/c2/c2_1994.pdf

Gadsen, J. and Outteridge, C. (2006). What Value Analysis: The Historical Record of Fratricide. 23rd International Symposium on Military Operational Research, August 29th–September 1st, 2006. Available at: http://ismor.cds.cranfield.ac.uk/ISMOR/2006/JGadsden.pdf

Gadsen, J., Krause, D., Dixon, M. and Lewis, L. (2008). Human Factors in combat ID – An international perspective. Human Factors Issues in Combat Identification, Gold Canyon Golf Resort, Arizona, USA, May 13th–14th. Available at: http://www.cerici.org/workshop/2008Workshops/Gadsden%20-%20CID%20Chapter.pdf

Genaidy, A., Salem, S., Karwoski, W., Paez, O. and Tuncel, S. (2007). The work compatibility improvement framework: An integrated perspective of the human at work system. *Ergonomics*, 50(1), 3–25.

Gertman, D. and Blackman, H.S. (1994). *Human Reliability and Safety Analysis Data Handbook*. New York: Wiley – IEEE.

Gibson, H., Walker, G., Stanton, N. and Baber, C. (2004). Report on results of EAST methodology for railway data: Possession scenario. HFI DTC Technical Report/WP1.1.3/7.

Gielo-Perczak, K. (2001). Systems approach to slips and falls. *Theoretical Issues in Ergonomics Science*, 12(2), 124–41.

Gillham, B. (2000). *Case Study Research Methods*. London: Continuum.

Glaser, B.G. and Strauss, A.L. (1967). *The Discovery of Grounded Theory.* Chicago, IL: Aldine.

Goosens, L. and Hale, A. (1997). Editorial: Risk assessment and accident analysis. *Safety Science,* 26(1/2), 21–3.

Gorman, J.C., Cooke, N.J. and Winner, J.L. (2006). Measuring team Situation Awareness in decentralized Command and Control environments. *Ergonomics,* 49(12), 1312–25.

Green, D.M. and Swets, J.A. (1966). *Signal Detection Theory and Psychophysics.* New York: John Wiley.

Greitzer, F.L. and Andrews, D.H. (2008). Training strategies to mitigate expectancy-induced response bias in Combat Identification: A research agenda. Human Factors Issues in Combat Identification, Gold Canyon Golf Resort, Arizona, USA, 13th–14th May. Available at: http://www.cerici.org

Greitzer, F.L. and Andrews, D.H. (2009). Training strategies to mitigate expectancy-induced response bias in Combat Identification: A research agenda, in *Human Factors in Combat Identification,* edited by D.H. Andrews, P.H. Robert and M.B. Wolf. Aldershot: Ashgate.

Griffin, T.G.C., Young, M.S. and Stanton, N.A. (2010). Investigating accident causation through information network modelling. *Ergonomics,* 53(2), 198–210.

Habraken, M.M.P., Van der Schaaf, T.W., Leistikow, I.P. and Reijnders-Thijssen, P.M.J. (2009). Prospective risk analysis of health care processes: A systematic evaluation of the use of HFMEA™ in Dutch health care. *Ergonomics,* 52(7), 809–19.

Hackman, J.R. (1987). The design of work teams, in *Handbook of Organizational Behaviour,* edited by J. Lorsch. Englewood Cliffs, NJ: Prentice Hall.

Hancock, P.A. (1997). Research Challenges in Human-Machine Systems. International Symposium on Human-Machine Systems, MIT: USA, December.

Harary, F. (1994). *Graph Theory.* Reading, MA: Addison-Wesley.

Hart, R. (2004). Fratricide: A dilemma which is manageable at best. Naval War College, Newport. Available at: http://www.dtic.mil/cgi-bin/GetTRDoc?Locat ion=U2&doc=GetTRDoc.pdf&AD=ADA422788

Hawley, J.K., Mares, A.L. and Marcon, J.L. (2009). On fratricide and the operational reliability of Target Identification decision aids in Combat Identification, in *Human Factors in Combat Identification,* edited by D.H. Andrews, P.H. Robert and M.B. Wolf. Aldershot: Ashgate.

Heinrich, H.W. (1931). *Industrial Accident Prevention.* New York: McGraw Hill.

Helmreich, R.L. (1984). Cockpit management attitudes. *Human Factors,* 26, 63–72.

Hendrick, K. and Benner, L. (1987). *Investigating Accidents with STEP.* New York: Marcel Dekker.

Heylighen, F. and Joslyn, C. (1992). What is Systems Theory? In Principia Cybernetica Web (Principia Cybernetica, Brussels) F. Heylighen, C. Joslyn and V. Turchin (eds) [Online]. Available at: http://pespmc1.vub.ac.be/SYSTHEOR.html

Hignett, S. (2005). Qualitative methodology in ergonomics, in *Evaluation of Human Work*, edited by J.R. Wilson and E. Megaw. London: Taylor and Francis.

Hinsz, F.L. and Andrews, D.H. (2009). Comparing individual and team judgement accuracy For Target Identification under heavy cognitive demand, in *Human Factors Issues in Combat Identification*, edited by D. Andrews, R.P. Hertz and M.B. Wolf. Aldershot: Ashgate.

Hinsz, F.L. and Wallace, D.M. (2009). A team training paradigm for better combat identification. In *Human Factors Issues in Combat Identificiation*, edited by Dee H. Andrews, Robert P. Herz and Mark B. Wolf. Aldershot: Ashgate.

Hirokawa, R.Y., Gouran, D.S. and Martz, A.E. (1988). Understanding the sources of faulty group decision making: A lesson from the *Challenger* disaster. *Small Group Research*, 19, 411–33.

Hirokawa, R.J. and Johnston, D.D. (1989). Toward a general theory of group decision making: Development of an integrated model. *Small Group Research*, 20, 500.

Hollingshead, A.B., Wittenbaum, G.M., Paulus, P.B., Hirokawa, R.Y., Ancona, D., Peterson, R.S., Jehn, K.A. and Yoon, K. (2005). A look at groups from the functional perspective, in *Theories of Small Groups: Interdisciplinary Perspectives*, edited by M.S. Poole and A.B. Hollingshead. Thousand Oaks, CA: Sage Publications.

Hollnagel, E. (1993). *Human Reliability Analysis – Context and Control*. London: Academic Press.

Hollnagel, E. (1998). *Cognitive Reliability and Error Analysis Method*. New York: Elsevier Science.

Hollnagel, E. (1999). Accident analysis and barrier functions. [Online]. Available at: http://www.it.uu.se/research/project/train/papers/AccidentAnalysis.pdf

Hollnagel, E. (2004). *Barriers and Accident Prevention*. Aldershot: Ashgate.

Hollnagel, E. (2005). Accident models and accident analysis. [Online]. Available at: http://www.ida.liu.se/~eriho/AccidentModels_M.htm

Hollnagel, E. (2007). Decisions about 'what' and decisions about 'how', in *Decision Making in Complex Environments*, edited by M.J. Cook, J.M. Noyes and Y. Masakowski. Aldershot: Ashgate.

Houghton, R.J., Baber, C., Cowton, M. and Stanton, N. (2008). WESTT (Workload, Error, Situational Awareness, Time and Teamwork): An analytical prototyping system for Command and Control. *Cognition, Technology and Work*, 10, 199–207.

Houghton, R.J., Baber, C., McMaster, R., Stanton, N.A., Salmon, P., Stewart, R. and Walker, G.H. (2006). Command and control in emergency services operations: A social network analysis. *Ergonomics*, 49(12), 1204–25.

Hourizi, R. and Johnson, P. (2003). Towards an explanatory, predictive account of awareness. *Computers and Graphics*, 27, 859–72.

Hutchins, E. (1995). How a cockpit remembers its speeds. *Cognitive Science*, 19, 265–88.

Hutchins, S.G., Hocevar, S.P. and Kemple, W.G. (1999). Analysis of team communications in 'human-in-the loop' experiments in Joint Command and Control. International Command and Control Research and Technology Symposium. Newport, R.I.,Naval War College, 29th June–1st July. Available at: http://www.dodccrp.org/html4/events_past.html#1999

Jamieson, G.A. and Wang, L. (2007). Developing human machine interfaces to support appropriate trust and reliance on automated Combat Identification Systems. Progress report for milestones 1, 2 and 3. University of Toronto. Available at: http://www.dtic.mil/cgi-bin/GetTRDoc?Location=U2&doc=Get TRDoc.pdf&AD=ADA485517

Jarmasz, J., Zobarich, R., Bruyn-Martin, L. and Lamoureux, T. (2009). Team cognition during a simulated Close Air Support exercise: Results from a new behavioural rating technique, in *Human Factors Issues In Combat Identification*, edited by D. Andrews, R.P. Hertz and M.B. Wolf. Aldershot: Ashgate.

Jarvis, S. and Harris, D. (2010). Development of a bespoke Human Factors taxonomy for gliding accident analysis and its revelations about highly inexperienced UK glider pilots. *Ergonomics*, 53(2), 294–303.

Jenkins, D.P., Salmon, P.M., Stanton, N.A. and Walker, G.H. (2010). A systemic approach to accident analysis: A case study of the Stockwell shooting. *Ergonomics*, 53(1), 1–17.

Jenkins, D.P., Salmon, P.M., Stanton, N.A., Walker, G.H. and Rafferty, L.A. (2011). What could they have been thinking? How sociotechnic system design influences cognition: A case study of the Stockwell shooting. *Ergonomics*, 54(2), 103–19.

Jenkins, D.P., Stanton, N.A., Walker, G.H., Salmon, P. and Young, M.S. (2008). Applying cognitive work analysis to the design of rapidly reconfigurable interfaces in complex networks. *Theoretical Issues in Ergonomics Science*, 9(4), 272–95.

Jensen, H.J. (2003). Foreword, in *Complexity Theory and Network Centric Warfare*, edited by J. Moffat. Washington, DC: CCRP Press.

Johnson, W.G. (1973). The Management Oversight and Risk Tree – MORT: Including systems developed by the Idaho Operations Office and Aerojet Nuclear Company. US Atomic Energy Commission Division of Operational Safety AT (04–3) B21. US Government Printing Office. Available at: http://nri.eu.com/SAN8212.pdf

Karwowski, W. (2006). *International Encyclopaedia of Ergonomics and Human Factors*. Bristol, PA: Taylor and Francis.

Kass, S.J., Herschler, D.A. and Companion, M.A. (1991). Training Situational Awareness through pattern recognition in a battlefield environment. *Military Psychology*, 3, 105–12.

Katsakiori, P., Sakellaropoulos, G. and Manatakis, E. (2009). Towards an evaluation of accident investigation methods in terms of their alignment within accident causation models. *Safety Science*, 47(7), 1007–15.

Katz, N., Lzaer, D., Arrow, H. and Contractor, N. (2005). The network perspective on small groups: Theory and research, in *Theories of Small Groups: Interdisciplinary Perspectives*, edited by M.S. Poole and A.B. Hollingshead. Thousand Oaks, CA: Sage Publications.

Kennedy, R. and Kirwan, B. (1998). Development of a hazard and operability-based method for identifying safety management vulnerabilities in high risk systems. *Safety Science*, 30, 249–74.

Kingston, J., Koornneef, F., van den Ruit, J., Frei, R. and Schallier, P. (2009). NRI MORT User's Manual. For use with the anagement Oversight and Risk Tree Analytical Logic Diagram. (2nd edition). Delft, The Netherlands: The Noordwijk Risk Initiative Foundation. Available at: http://nri.eu.com/NRI1.pdf

Kingston, J., Nertney, R., Frei, R. and Schallier, P. (2004). Barrier Analysis analysed in MORT Perspective. *PSAM7ESREL*, 4, 364–9. Available at: http://nri.eu.com/PSAM-FINAL.pdf

Kirwan, B. (1992a). Human error identification in human reliability assessment. Part 1: Overview of approaches. *Applied Ergonomics*, 23(5), 299–318.

Kirwan, B. (1992b). Human errorsIdentification in human reliability assessment. Part 2: Detailed comparison of techniques. *Applied Ergonomics*, 23(6), 371–81.

Kirwan, B. (1994). *A Guide to Practical Human Reliability Assessment*. London: Taylor and Francis.

Kirwan, B. (1996). Reliability quantification techniques – THERP, HEART and JHEDI: Part 1 – Technique descriptions and validation issues. *Applied Ergonomics*, 27(6), 359–73.

Kirwan, B. (1998a). Human error identification techniques for risk assessment of high risk systems – Part 1: Review and evaluation of techniques. *Applied Ergonomics*, 29(3), 157–77.

Kirwan, B. (1998b). Human error identification techniques for risk assessment of high risk systems – Part 2: Towards a framework approach. *Applied Ergonomics*, 29(5), 299–318.

Kirwan, B. and Ainsworth, L.K. (1992). *A Guide to Task Analysis*. London: Taylor and Francis.

Klein, G. (1989). Recognition-primed decisions. *Advances in Man-Machine Systems Research*, 5, 47–92.

Klein, G. (1997). Current status of NDM, in *Decision Making Under Stress*, edited by R. Flin, E. Salas, M. Strub and L. Martin. Aldershot: Ashgate.

Klein, G. (2003). *The Power of Intuition: How to Use Your Gut Feelings to Make Better Decisions at Work*. New York: Currency.

Klein, G. (2008). Naturalistic decision making. *Human Factors*, 50(3), 456–60.

Klein, G. and Armstrong, A.A. (2005). Critical decision method, in *Handbook of Human Factors and Ergonomics Methods*, edited by N.A. Stanton et al. London: CRC Press.

Klein, G. and Hoffman, R. (2008). Macrocognition, mental models and Cognitive Task Analysis methodology, in *Naturalistic Decision Making and Macrocognition*, edited by J.M. Schraagen, L.G. Militello, T. Ormerod and R. Lipshitz. Aldershot: Ashgate.

Klimoski, R., and Mohammed, S. (1994). Team mental model: Construct or metaphor? *Journal of Management*, 20, 403–37.

Kogler, T.M. (2003). The effects of degraded vision and automatic Combat Identification reliability on infantry friendly fire engagements. Master's thesis. Petersburg, PA: Virginia Polytechnic Institute and State University. Available at: http://scholar.lib.vt.edu/theses/available/etd-03272003-165934/unrestricted/ThesisEDTversion.pdf

Korsgaard, M.A., Brodt, S.E. and Sapienza, H.J. (2005). Trust, identity and attachment: Promoting individuals' cooperation in groups, in *The Essentials of Teamworking: International Perspectives*, edited by M.A. West, D. Tjosvold and K.G. Smith. Chichester: John Wiley and Sons Ltd.

Ladkin, P.B. and Stuphorn, J. (2003). Two causal analyses of the Black Hawk shootdown during Operation Provide Comfort. Eighth Australian Workshop on Safety Critical Systems and Software (SCS 2003), Canberra, Australia. CRPIT, 33. Edited by P. Lindsay and T. Cant. ACS. 3–23. Available at: http://crpit.com/abstracts/CRPITV33Ladkin.html

Laflamme, L. (1990). A better understanding of occupational accident genesis to improve safety in the workplace. *Journal of Occupational Accidents*, 12, 155–65.

Langan-Fox, J., Code, S. and Langfield Smith, K. (2000). Team mental models: Techniques, methods and analytic approaches. *Human Factors*, 42(2), 242–71.

Langan-Fox, J., Wirth, A.I., Code, S.L., Langfield-Smith, K. and Wirth, A. (2001). Analysing shared and team mental models. *International Journal of Industrial Ergonomics*, 28, 99–112.

Lehto, M. and Salvendy, G. (1991). Models of accident causation and their application: Review and reappraisal. *Journal of Engineering and Technical Management*, 8, 173–205.

Leveson, N. (2001). Evaluating accident models using recent aerospace accidents. NASA Report. Available at: http://sunnyday.mit.edu/accidents/nasareport.pdf

Leveson, N. (2002). *A New Approach to System Safety Engineering*. Cambridge, MA: Aeronautics and Astronautics, Massachusetts Institute of Technology.

Leveson, N. (2004). A new accident model for engineering safer systems. *Safety Science*, 42, 237–70.

Leveson, N., Allen, P. and Storey, M.-A. (2002). The analysis of a friendly fire accident using a systems model of accidents. 20th International System Safety Society Conference (ISSC 2003). System Safety Society, Unionville, Virginia. Available at: http://citeseerx.ist.psu.edu

Leximancer (2009). Leximancer Manual version 3.1. Leximancer Pty, Ltd.

Li, W.C., Harris, D. and Yu, C.S. (2008). Routes to failure: Analysis of 41 civil aviation accidents from the Republic of China using the Human Factors

Analysis and Classification System. *Accident Analysis and Prevention*, 40, 426–34.

MacMillan, J., Paley, M.J., Levchuk, Y.N., Entin, E.E., Serfaty, D. and Freeman, J.T. (2002). Designing the best team for the task: Optimal organizational structures for military missions, in *New Trends in Cooperative Activities: System Dynamics in Complex Settings*, edited by M. McNeese, E. Salas, and M. Endsley. San Diego, CA: Human Factors and Ergonomics Society Press.

Manning, F.J. (1991). Morale, cohesion, and esprit de corps, in *Handbook of Military Psychology*, edited by R. Gal and A.D. Mangelsdorff. New York: John Wiley.

Markovsky, B., Heimer, K. & O'Brien, J. O. (1994). *Advances in group processes*. Greenwich, CT: JAI.

Marques, J. and McCall, C. (2005). The application of interrater reliability as a solidification instrument in a phenomenological study. *The Qualitative Report*, 10(3), 439–62.

Masys, A.J. (2006). Understanding fratricide: Insights from actor network theory and complexity theory. International System Safety Conference, Albuquerque, New Mexico, 31st July–4th August 2006.

Maule, A.J. (2010). Can computers overcome limitations in human decision making? Paper delivered at the NDM09 Conference, 23rd–26th June 2009, London, reprinted in *International Journal of Human Computer Interaction*, 26, 108–19.

McCann, C., Baranski, J.V., Thompson, M.M. and Pigeau, R.A. (2000). On the utility of experiential cross-training for team decision making under time stress. *Ergonomics*, 43(8), 1095–110.

McIntyre, R.M. and Dickinson, T.L. (1992). Systemic assessment of teamwork processes in tactical environments. Report for Naval Training Systems Center, contract No. N61339-91-C-0145). Norfolk, VA: Old Dominion University.

Merriam-Webster (2011). Call sign. [Online]. Available at: http://www.merriam-webster.com/dictionary/call%20sign

Militello, L.G., Kyne, M.M, Klein, G., Getchell, K. and Thordsen, M. (1999). A synthesized model of team performance. *International Journal of Cognitive Ergonomics*, 3(2), 131–58.

Ministry of Defence (2002). *Combat Identification:* Report by the Comptroller and Auditor General. HC661 Session 2001–2002. London: The Stationary Office. Available at: http://www.nao.org.uk/publications/0102/mod_combat_identification.aspx

Ministry of Defence (2003*). United Kingdom Glossary of Joint and Multinational Terms and Definitions.* Joint Doctrine Publication 0–01.1. Seventh edition. Available at: http://www.mod.uk/NR/rdonlyres/E8750509-B7D1-4BC6-8AE E-8A4868E2DA21/0/JDP0011Ed7.pdf

Ministry of Defence (2004). Board of Inquiry into the Challenger 2 Incident – 25 March 2003. Available at: http://www.mod.uk/DefenceInternet/About

Defence/CorporatePublications/BoardsOfInquiry/BoardOfInquiryInto
TheChallenger2Incident25Mar03.htm

Ministry of Defence, Defence Concepts and Doctrine Centre (2008). Strategic Trends Programme, Future Character of Conflict. London: MoD (DCDC) Forms and Publications Section. Available at: http://www.mod.uk/DefenceInternet/ MicroSite/DCDC/OurPublications/Concepts/FutureCharacterOfConflict.htm (accessed 16 February 2012).

Mistry, B., Croft, G., Dean, D., Gadsen, J., Conway, G. and Cornes, K. (2009). Analysis of the tasks conducted by forward air controllers and pilots during simulated Close Air Support missions: Supporting the development of the INCIDER Model, in *Human Factors in Combat Identification*, edited by D.H. Andrews, P.H. Robert and M.B. Wolf. Aldershot: Ashgate.

Mitchell, L. and Flin, R. (2007). Decisions to shoot by police officers. *Journal of Cognitive Engineering and Decision Making*, 1, 375–90.

Modarres, M. (1993). *What Every Engineer Should Know About Reliability and Risk Analysis*. New York: CRC Press/Marcel Dekker Ltd.

Moffat, J. (2003a). Quantifying the benefit of collaboration across an information network. *Journal of Defence Sciences*, 8(3), 23–128.

Moffat, J. (2003b). *Complexity Theory and Network Centric Warfare*. Washington, DC: CCRP Press.

Monge, P.R. and Contractor, N.S. (2001). Emergence of communication networks, in *Handbook of Organisational Communication*, edited by F.M. Jablin and L.L. Putnam. Thousand Oaks, CA: Sage.

Monge, P.R. and Contractor, N.S. (2003). *Theories of Communications Networks*. Oxford: Oxford University Press.

Moore, R.A., Schermerhorn, J.H., Oonk, H.M. and Morrison, J.G. (2003). Understanding and improving knowledge transactions in Command and Control. Report by Pacific Science and Engineering Group. Available at: http://www.dtic.mil/cgi-bin/GetTRDoc?Location=U2&doc=GetTRDoc. pdf&AD=ADA467096

Mosier, K.L. (2008). Technology and naturalistic decision making: Myths and realities, in *Naturalistic Decision Making and Macrocognition*, edited by M. Schraagen, L.G. Militello, T. Ormerod and R. Lipshitz. Aldershot: Ashgate.

Naikar, N. and Saunders, A. (2002). Crossing the boundaries of safe operation: Training for error detection and error recovery, in Workshop on the Investigation and Reporting of Incidents and Accidents (IRIA), Glasgow, 17th–20th July 2002, edited by C.W. Johnson. Available at: http://www.dcs. gla.ac.uk/~johnson/iria2002/IRIA_2002.pdf

Neisser, U. (1976). *Cognition and Reality*. San Francisco, CA: Freeman.

Neyedli, H.F., Wang, L., Jamieson, G.A. and Hollands, J.G. (2009). Evaluating reliance on Combat Identification systems: The role of reliability feedback, in *Human Factors in Combat Identification*, edited by D.H. Andrews, P.H. Robert and M.B. Wolf. Aldershot: Ashgate.

Norman, D.A. (1981). Categorization of action slips. *Psychological Review*, 88(1), 1–15.

Numagami, T. (1998). The infeasibility of invariant laws in management studies: A reflective dialogue in defense of case studies. *Organisation Science*, 9(1), 2–15.

O'Connor, P., O'Dea, K. and Melton, J. (2007). A methodology for identifying human error in US Navy diving accidents. *Human Factors*, 49(2), 214–26.

O'Connor, P., O'Dea, A., Flin, R. and Belton, S. (2008). Identifying the team skills required by nuclear power plant operations personnel. *International Journal of Industrial Ergonomics*, 38, 1028–37.

Ogden, G.C. (1987). Concept, knowledge and thought. *Annual Review of Psychology*, 38, 203–27.

Omodei, M., Wearing, A. and McLennan, J. (2000). Relative efficacy of an open versus a restricted communication structure for Command and Control decision making: An experimental study, in *The Human in Command: Exploring the Modern Military Experience*, edited by C. McCann and R. Pigeau. New York: Kluwer Academic/ Plenum Publishers.

Ong, Yu Lin, and Lim, Beng Chong (2004). Decision making in a Brigade Command Team: Integrating theory and practice. *Pointer Journals*, 30(4).

Orasanu, J.M. (1995). Evaluating Team Situation Awareness through Communication, in Proceedings of the International Conference on Experimental Analysis and Measurement of Situation Awareness, Daytona Beach, FL, USA, edited by D. Garland and M. Endsley.

Orasanu, J.M. (2005). Crew collaboration in space: A naturalistic decision making perspective. *Aviation, Space and Environmental Medicine*, 76(6), Section II.

Orasanu, J.M. and Salas, E. (1993). Team decision making in complex environments, in *Decision-making in Action: Models and Methods*, edited by G.A. Klein, J. Orasanu, R. Calderwood and C.E. Zsambok. Norwood, NJ: Ablex.

Oxford English Dictionary (2010). *Deconfliction*. [Online.] Available at: www.oed.com

Oxforddictionaries (2011). *Battle Group*. Oxford University Press. [Online.] Available at: www.oxforddictionaries.com

Pae, P. (2003). Friendly Fire Still a Problem. *The Los Angeles Times*. [Online]. Available at: http://www.globalsecurity.org/org/news/2003/030516-friendly-fire01.htm

Parasuraman, R. and Riley, V. (1997). Humans and automation: Use, misuse, disuse, abuse. *Human Factors*, 39.

Paries, J. (2006). Complexity, emergence and resilience, in *Resilience Engineering: Concepts and Precepts*, edited by E. Hollnagel, D.W. Woods and N. Leveson. Aldershot: Ashgate.

Paris, C.R., Salas, E. and Cannon-Bowers, J.A. (2000). Teamwork in multi-person systems: A review and analysis. *Ergonomics*, 43(8), 1052–75.

Passmore, W., Francis, C., Haldeman, J. and Shani, A. (1982). Sociotechnical systems: A North American reflection on empirical studies of the seventies. *Human Relations*, 35(12), 1179–204.

Patrick, J., James, N. and Ahmed, A. (2006). Human processes of control: Tracing the goals and strategies of control room teams. *Ergonomics*, 49(12), 1395–414.

Patrick, J., James, N., Ahmed, A. and Halliday, P. (2006). Observational assessment of Situation Awareness, team differences and training implications. *Ergonomics*, 49(4), 393–417.

Pharaon, J.W. (2009). Mitigating friendly fire casualties though enhanced battle command capabilities, in *Human Factors in Combat Identification*, edited by D.H. Andrews, P.H. Robert and M.B. Wolf. Aldershot: Ashgate.

Pidgeon, N. and Henwood, K. (1997). Using grounded theory in Psychological research, in *Doing Qualitative Analysis in Psychology*, edited by N. Hayes. Hove: Psychology Press.

Pigeau, R. and McCann, C. (2000). Redefining command and control, in *The Human in Command: Exploring the Modern Military Experience*, edited by C. McCann and R. Pigeau. New York: Kluwer Academic/Plenum Publishers.

Piper, J.L. (2001). *A Chain of Events: The Government Cover-Up of the Black Hawk Incident and the Friendly-Fire Death of Lt. Laura Piper*. Virginia: Brasseys, Inc.

Pirnie, B.R., Vick, A.J., Grissom, A., Mueller, K.P. and Orletsky, D.T. (2005). Beyond Close Air Support. Forging a new air–ground partnership. Prepared for the United States Air Force by RAND. Available at: www.rand.org/pubs/monographs/2005/RAND_MG301.pdf

Pretorious, A. and Cilliers, P.J. (2007). Development of a mental workload index: A systems approach. *Ergonomics*, 50(9), 1503–15.

Qureshi, Z. (2006). Safety in sociotechnical systems: Towards a multi-disciplinary approach. [Online]. Available at: http://www.cedisc.com/safety/Sociotechnical.pps

Qureshi, Z. (2007). A Review of Accident Modelling Approaches for Complex Socio-Technical Systems. 12th Australian Conference on Safety-Related Programmable Systems: Adelaide, Australia. Available at: http://crpit.com/confpapers/CRPITV86Qureshi.pdf

Rasker, P.C., Post, W.M. and Schraagen, J.M. (2000). Effects of two types of intra-team feedback on developing a shared mental model in Command and Control teams. *Ergonomics*, 43(8), 1167–89.

Rasmussen, J. (1997). Risk management in a dynamic society: A modelling problem. *Safety Science*, 27(2/3), 183–213.

Rasmussen, J. (2000). Human Factors in a dynamic information society: Where are we heading? *Ergonomics*, 43(7), 869–79.

Rasmussen, J. and Svedung, I. (2000). *Proactive Risk Management in a Dynamic Society*. Swedish Rescue Agency. Sjuhardsbygdens Tryckert, Boras.

Reason, J. (1990). *Human Error*. Cambridge: Cambridge University Press

Regan, G. (2004). *More Military Blunders*. London: Carlton Books.

Rentsch, J.R. and Hall, R.J. (1994). Members of great teams think alike: A model of team effectiveness and schema similarity among team members. In *Advances in Interdisciplinary Studies of Work Teams: Theories of Self-Managing Work Teams*, edited by M.M. Beyerlein and D.A. Johnson. Greenwich, CT: JAI Press.

Rice, S., Clayton, K. and McCarley, J. (2009). The effects of automation bias on operator compliance and reliance, in *Human Factors in Combat Identification*, edited by D.H. Andrews, P.H. Robert and M.B. Wolf. Aldershot: Ashgate.

Rouse, W.B., Cannon-Bowers, J.A. and Salas, E. (1992). The role of mental models in team performance in complex systems. *IEEE Transactions on Systems, Man and Cybernetics*, 22, 1296–308.

Salas, E., Cooke, N.J. and Rosen, M.A. (2008). On teams, teamwork, and team performance: discoveries and developments. *Human Factors*, 50, 540–47

Salas, E., Muniz, E.J. and Prince, C. (2006). Situation Awareness in teams, in *International Encyclopaedia of Ergonomics and Human Factors*, edited by W. Karwowski. Bristol, PA: Taylor and Francis.

Salas, E., Prince, C., Baker, D. and Shrestha, L. (1995). Situation Awareness in team performance: Implications for measurement and training. *Human Factors*, 37(1), 123–36.

Salas, E., Rosen, M.A., Burke, C.S., Nicholson, D. and Howse, W.R. (2007). Markers for enhancing team cognition in complex environments: The power of team performance diagnosis. *Aviation, Space and Environmental Medicine*, 78(5), Section II.

Salas, E., Sims, D.E., and Burke, C.S. (2005). Is there a big five in teamwork? *Small Group Research*, 36(5), 555–99.

Salmon, P., Regan, M. and Johnston, I. (2005). Human error and road transport: Phase 1 – Literature review. ATSB Final Report. Monash University Accident Research Centre.

Salmon, P.M., Stanton, N.A., Walker, G.H. and Jenkins, D.P. (2009). *Distributed Situation Awareness: Advances in Theory, Measurement and Application to Teamwork*. Aldershot: Ashgate.

Salmon, P.M., Stanton, N.A., Walker, G., Baber, C., Jenkins, D. and McMaster, R. (2008). Representing Situation Awareness in collaborative systems: A case study in the energy distribution domain. *Ergonomics*, 51(3), 367–84.

Salmon, P.M., Stanton, N.A., Walker, G.H., Baber, C., Jenkins, D.P., McMaster, R. and Young, M.S. (2008). What is really going on? Review of Situation Awareness models for individuals and teams. *Theoretical Issues in Ergonomics Science*, 9(4), 297–323.

Salmon, P.A., Stanton, N.A., Young, M.S., Harris, D., Demagalski, J., Marshall, A., Waldmann, T. and Dekker, S. (2002). *Using Existing HEI Techniques to Predict Pilot Error: A Comparison of SHERPA, HAZOP and HEIST*. In Proceedings of International Conference on Human Computer Interaction in Aeronautics – HCI – Aero 2002, Menlo Park, CA. Edited by S. Chatty, J. Hansman and G. Boy. AAAI Press.

Salmon, P., Walker, G. and Stanton, N.A. (2004). National grid Transco: Alarm handling scenario. HFI DTC Technical Report/WP 1.1.3.

Schraagen, J.M., Militello, L.G., Ormerod, T. and Lipshitz, R. (2008). *Naturalistic Decision Making and Macrocognition*. Aldershot: Ashgate

Sebok, A. (2000). Team performance in process control: Influences of interface design and staffing levels. *Ergonomics*, 43(8), 1210–36.

Serfaty, D., Entin, E. and Volpe, C. (1993). Adaptation to stress in team decision making and coordination. Human Factors and Ergonomics Society 37th Annual Meeting, Washington State Convention Center, USA, 11th–15th October 1993.

Shahrokhi, M. and Bernard, A. (2008). Energy flow/barrier analysis, a novel view. [Online]. Available at: http://www.univvalenciennes.fr/congres/EAM06/PDF_Papers_author/Session8_Shahrokhi.pdf

Shanteau, J. (1992). How much information does an expert use? Is it relevant? *Acta Psychologica*, 81, 75–86.

Shanteau, J. and Stewart, T.R. (1992). Why study expert decision making? Some historical perspectives and comments. *Organisational Behaviour and Human Decision Process*, 53, 95–106.

Shappell, S.A. and Wiegmann, D.A. (2000). The Human Factors Analysis and Classification System (HFACS) (Report Number DOT/FAA/AM-00/7). Washington, DC: Office of Aviation Medicine. Available at: http://www.nifc.gov/fireInfo/fireInfo_documents/humanfactors_classAnly.pdf

Shappell, S., Detwieler, C., Holcomb, K., Hackworth, C., Boquet, A. and Wiegmann, D.A. (2007). Human error and commercial aviation accidents: An analysis using the Human Factors Analysis and Classification System. *Human Factors*, 49(2), 227–42.

Shattuck, L.G. and Woods, D.D. (2000). Communication of intent in military command and control systems, in *The Human in Command: Exploring the Modern Military Experience*, edited by C. McCann and R. Pigeau. New York: Kluwer Academic/Plenum Publishers.

Shebilske, W., Levchuk, G., Freeman, J. and Gildea, K. (2009). A team training paradigm for better Combat Identification, in *Human Factors in Combat Identification*, edited by D.H. Andrews, P.H. Robert and M.B. Wolf. Aldershot: Ashgate.

Shorrock, S.T. (2002). Error classification for safety management: finding the right approach, in Workshop on the Investigation and Reporting of Incidents and Accidents (IRIA), Glasgow, 17th–20th July 2002, edited by C.W. Johnson. Available at: http://www.dcs.gla.ac.uk/~johnson/iria2002/IRIA_2002.pdf

Shorrock, S.T. and Kirwan, B. (2002). Development and application of a human error identification tool for Air Traffic Control. *Applied Ergonomics*, 33, 319–336.

Shrader, C.R. (1982). *Amicide: The Problem of Friendly Fire in Modern War.* Fort Leavenworth, KS: Combat Studies Institute.

Siegel, A.I. and Federman, P.J. (1973). Communications content training as an ingredient in effective team performance. *Ergonomics*, 16(4), 1366–5847.

Sklet, S. (2002). Methods for accident investigation reliability, safety and security studies at NTNU – Norwegian University Of Science And Technology. Available at: http://www.ipk.ntnu.no/ross

Sklet, S. (2004). Comparison of some selected methods for accident investigation. *Journal of Hazardous Materials*, 111, 29–37.

Smith, A.E. and Humphreys, M.S. (2006). Evaluation of unsupervised semantic mapping of natural language with Leximancer concept mapping. *Behaviour Research Methods*, 38(2), 262–79.

Smith, K. and Hancock, P.A. (1995). Situation awareness is adaptive, externally directed consciousness. *Human Factors*, 37, 137–48.

Smith, W. and Dowell, J. (2000). A case study of coordinative decision making in disaster management. *Ergonomics*, 43(8), 1153–66.

Snook, S.A. (2000). *Fratricide: The Accidental Shootdown of US Black Hawks Over Northern Iraq*. Princeton, NJ: Princeton University Press.

Soukas, J. (1988). The role of safety analysis in accident prevention. *Accident Analysis and Prevention*, 20(1), 67–85.

Stagl, K.C., Burke, C.S., Salas, E., and Peirce, L. (2006). Team adaptation: Realizing team synergy, in C.S. Burke, L.G. Pierce, and E. Salas (eds), *Understanding Adaptability: A Prerequisite for Effective Performance Within Complex Environments*. Oxford: Elsevier, 117–41.

Stanton, N.A. (2006). Hierarchical task analysis: Developments, applications and extensions. *Applied Ergonomics*, 37, 55–79.

Stanton, N.A. and Ashleigh, M.J. (2000). A field study of teamworking in a new human supervisory control system. *Ergonomics*, 43(8), 1190–209.

Stanton, N.A. and Baber, C. (1996). A systems approach to human error identification. *Safety Science*, 22(13), 215–28.

Stanton, N.A. and Baber, C. (2005). Validating TAFEI: Reliability and validity of a human error prediction technique. *Ergonomics*, 48(9), 1097–113.

Stanton, N.A. and Baber, C. (2008). Modelling of human alarm handling response times: A case study of the Ladbroke Grove rail accident in the UK. *Ergonomics*, 51(4), 423–40.

Stanton, N.A. and Young, M.S. (1999). What price Ergonomics? *Nature*, 399, 197–8.

Stanton, N.A., Baber, C. and Harris, D. (2008). *Modelling Command and Control: Event Analysis of Systemic Teamwork*. Aldershot: Ashgate.

Stanton, N.A., Baber, C., Walker, G.H., Houghton, R.J., McMaster, R., Stewart R., Harrison, A.G., Jenkins, D.P., Young, M.S. and Salmon, P.M. (2008). Development of a generic activities model of command and control. *Cognition, Technology and Work*, 10(3), 209–20.

Stanton, N.A., Chambers, P.R.G. and Piggot, J. (2001). Situation awareness and safety. *Safety Science*, 39, 189–204.

Stanton, N.A., Jenkins, D.P., Salmon, P.M., Walker, G.H., Revell, K.M.A. and Rafferty, L.A. (2009). *Digitising Command and Control: A Human Factors and Ergonomics Analysis of Mission Planning and Battlespace Management.* Aldershot: Ashgate.

Stanton, N.A., Rafferty, L.A., Salmon, P.M., Revell, K.M., McMaster, R., Caird-Daley, A. and Cooper-Chapman, C. (2010). Distributed decision making in multi-helicopter teams: A case study of mission planning and execution from a non-combatant evacuation operation training scenario. *Journal of Cognitive Engineering and Decision Making*, 4(4), 328–53(26).

Stanton, N.A., Salmon, P.M., Walker, G.H. and Jenkins, D.P. (2009a). Genotype and phenotype schemata and their role in Distributed Situation Awareness in collaborative systems. *Theoretical Issues in Ergonomics Science*, 10(1), 43–68.

Stanton, N.A., Salmon, P.M., Walker, G.H. and Jenkins, D.P. (2009b). Is situation awareness all in the mind? *Theoretical Issues in Ergonomics Science*, 11(1), 29–40.

Stanton, N.A., Salmon, P.M., Walker, G.H., Baber, C. and Jenkins, D.P. (2005). *Human Factors Methods: A Practical Guide for Engineering and Design.* Ashgate: Aldershot.

Stanton, N.A., Stewart, R., Harris, D., Houghton, R.J., Baber, C., McMaster, R., Salmon, P., Hoyle, G., Walker, G., Young, M.S., Linsell, M., Dymott, R. and Green, D. (2006). Distributed situation awareness in dynamic systems: Theoretical development and application of an ergonomics methodology. *Ergonomics*, 49(12–13), 1288–311.

Stanton, N.A., Baber, C., Walker, G.H., Houghton, R.J., McMaster, R., Stewart, R., Harris, D., Jenkins, D.P., Young, M.S. and Salmon, P.M. (2008) Development of a generic activities model of command and control. *Cognition, Technology and Work*, 10 (3), 209–20.

Steinweg, K.K. (1995). Dealing realistically with fratricide. *Parameters*, Spring, 4–29. Available at: http://www.carlisle.army.mil/usawc/Parameters/Articles/1995/steinweg.htm

Stewart, R.J., Stanton, N.A., Harris, D., Baber, C., Salmon, P., Mock, M., Tatlock, K., Wells, L. and Kay, A. (2008). Distributed Situational Awareness in an Airborne Warning and Control Aircraft: Application of a novel Ergonomics methodology. *Cognition, Technology and Work*, 10, 221–9.

Stout, R.J. and Salas, E. (1993). *The Role of Planning in Coordinated Team Decision Making: Implications for Training.* Human Factors and Ergonomics Society 37th Annual Meeting, Washington State Convention Center, USA, 11th–15th October 1993.

Stout, R.J. Cannon-Bowers, J.A., and Salas, E. (1996). The role of Shared Mental Models in developing Team Situation Awareness: Implications for training. *Training Research Journal*, 2, 85–116.

Stout, R.J., Cannon-Bowers, J.A., Salas, E. and Milanovich, D.M. (1999). Planning, Shared Mental Models and coordinated performance: An empirical link is established. *Human Factors*, 41(1), 61–71.

Strauss, A. and Corbin, J. (1998). *Basics of Qualitative Research: Techniques and Procedures for Developing Grounded Theory*. Thousand Oaks, CA: Sage Publications.

The *Sun* (2009). 2nd 'friendly fire' death in 24hrs., December 23.

Svedung, I. and Rasmussen, J. (2002). Graphic representation of accident scenarios: Mapping system structure and the causation of accidents. *Safety Science*, 40, 397–417.

Svensson, J. and Andersson, J. (2006). Speech acts, communication problems, and fighter pilot team performance. *Ergonomics*, 49(12–13), 1226–37.

Swain, A.D. (1987). Accident sequence evaluation program human reliability analysis procedure. United States Department of Energy. US Nuclear Regulatory Commission. NUREG/CR-4772 SAND86-1996. Available at: http://www.osti.gov/bridge/servlets/purl/6370593-0eXswa/6370593.pdf

Tappin, D.C., Bentley, T.A. and Vitalis, A. (2008). The role of contextual factors for musculoskeletal disorders in the New Zealand meat processing industry. *Ergonomics*, 51(10), 1576–93.

Trist, E.L. (1978). On socio-technical systems, in *Sociotechnical Systems: A Sourcebook*, edited by W.A. Pasmore and J.J. Sherwood. San Diego, CA: University Associates.

Urban, J.M., Bowers, C.A., Monday, S.D. and Morgan Jr, B.B. (1995). Workload, team structure, and communication in team performance. *Military Psychology*, 7(2), 123–39.

Urban, J.M., Weaver, J.L., Bowers, C.A. and Rhodenizer, L. (1996). Effects of workload and structure on team processes and performance: Implications for complex team decision making. *Human Factors*, 38(2), 300–310.

USAF Aircraft Accident Investigation Board (1994). US Army Black Hawk helicopters 87-26000 and 88-26060: Vol. 1, Executive Summary: UH-60 Black Hawk Helicopter Accident, 14 April 1994. Available at: http://www.dod.gov/pubs/foi/reading_room/973-1.pdf.

US Congress, Office of Technology Assessment (1993). Who goes there: Friend or foe? OTA – ISC-537. Washington, DC: US Government Printing Office. Available at: http://www.fas.org/ota/reports/9351.pdf

US Government Accountability Office (1997). Office of Special Investigations. Operation Provide Comfort: Review of Air Force investigation of Black Hawk fratricide incident (GAO/OSI-98-4). Washington, DC: US Government Printing Office.

Vicente, K.J., and Christoffersen, K. (2006). The Walkerton *E. coli* outbreak: A test of Rasmussens' framework for risk management in a dynamic society. *Theoretical Issues in Ergonomics Science*, 7(2), 92–112.

Von Bertalanffy, L. (1950). An outline of general system theory. *The British Journal for the Philosophy of Science*, 1(2), 134–65.

Wagenaar, W.A. and Schrier, J.V. (1997). Accident analysis the goal, and how to get there. *Safety Science*, 26(1/2), 25–33.

Walker, G.H., Gibson, H., Stanton, N.A., Baber, C., Salmon, P. and Green, D. (2006). Event Analysis of Systemic Teamwork (EAST): A novel integration of Ergonomics methods to analyse C4i activity. *Ergonomics*, 49, 1345–1369.

Walker, G.H., Stanton, N.A. and Salmon, P.M. (2011). Cognitive compatibility of motorcylists and car drivers. *Accident Analysis and Prevention*: Special Issue on Powered Two Wheelers Inside the Traffic System.

Walker, G.H., Stanton, N.A., Baber, C., Wells, L., Gibson, H., Salmon, P. and Jenkins, D. (2010). From Ethnography to the EAST method: A tractable approach for representing distributed cognition in Air Traffic Control. *Ergonomics*, 53(2), 184–97.

Walker, G.H., Stanton, N.A., Jenkins, D.P., Salmon, P.M. and Rafferty, L.A. (2009). From the 6Ps of Planning to the 4Ds of Digitisation: Difficulties, Dilemmas and Defective Decision making. NDM09 Conference, 23–26 June 2009, London UK.

Walker, G.H., Stanton, N.A., Jenkins, D.P., Salmon, P.M. and Rafferty, L.A. (2010). From the 6Ps of planning to the 4 Ds of digitization: Difficulties, dilemmas, and defective decision making. *International Journal of Human Computer Interaction*, 26(2), 173–88.

Walker, G.H., Stanton, N.A., Kazi, T.A., Salmon, P.M. and Jenkins, D.P. (2009). Does advanced driver training improve Situation Awareness? *Applied Ergonomics*, 40, 678–87.

Walker, G.H., Stanton, N.A., Salmon, P.M. and Jenkins, D.P. (2008). A review of sociotechnical systems theory: A classic concept for new Command and Control paradigms. *Theoretical Issues in Ergonomics Science*, 9(6), 479–99.

Walker, G.H., Stanton, N.A., Salmon, P.M. and Jenkins, D.P. (2009). *Command and Control: The Sociotechnical Perspective*. Aldershot: Ashgate.

Walker, G.H., Stanton, N.A., Salmon, P.M., Jenkins, D.P., Rafferty, L.A. and Ladva, D. (2010). Same or different? Generalising from novices to experts in Military Command and Control studies. *International Journal of Industrial Ergonomics*, 40, 473–83.

Walsh, J.P., Henderson, C.M. and Deighton, J. (1988). Negotiated belief structures and decision performance: An empirical investigation. *Organizational Behavior and Human Decision Processes*, 42, 194–216.

Walsham, G. (1995). Interpretive case studies in IS research: Nature and method. *European Journal of Information Systems*, 4(2), 74–81.

Wang, W.P., Luh, P.B., Serfaty, D. and Kleinman, D. (1991). Hierarchical team coordination in dynamic decision-making, in IEEE International Conference on Systems, Man, and Cybernetics, 13th–16th October 1991, Decision-aiding for Complex Systems. Available from IEEE Digital Library.

Watts, L.A. and Monk, A.F. (2000). Reasoning about tasks, activities and technology to support collaboration, in *Task Analysis*, edited by J. Annett and N. Stanton. London: Taylor and Francis.

Webber, S.S., Chen, G., Payne, S.C., Marsh, S.M. and Zaccaro, S.J. (2000). Enhancing team mental model measurement with performance appraisal practices. *Organisational Research Methods*, 3(4), 307–22.

Weisstein, E.W. (2008). Graph Diameter. From Math World – A Wolfram web resource. [Online]. Available at: http://www.mathworld.wolfram.com/Graph Diameter.html

West, P. (2007). We Brits invented friendly fire. Spiked. [Online]. Available at: http://www.spiked-online.com/index.php?/site/article/3774/

Wickens, C.D., Hyman, F., Dellinger, J., Taylor H. and Meador, M. (1986). The Sternberg memory search task as an index of pilot workload. *Ergonomics*, 29, 1371–83.

Wiegmann, D.A. and Shappell, S.A. (2001a). A human error analysis of commercial aviation accidents using the Human Factors Analysis and Classification System (HFACS). US Department of Transportation Federal Aviation Administration: Final Report. Available at: http://www.faa.gov/library/reports/medical/oam techreports/2000s/media/0103.pdf

Wiegmann, D.A. and Shappell, S.A. (2001b). Human error analysis of commercial aviation accidents: Application of the Human Factors Analysis and Classification System (HFACS). *Aviation, Space, and Environmental Medicine*, 72, 1006–16.

Wiegmann, D.A. and Shappell, S.A. (2003). *A Human Error Approach to Aviation Accident Analysis: The Human Factors Analysis and Classification System*. Aldershot: Ashgate.

Wiegmann, D.A. and Shappell, S.A. (2004). *HFACS* Analysis of military and civilian aviation accidents: A North American comparison. [Online]. Available at: http://www.asasi.org/2004_PPTs/Shappell%20et%20al_HFACS_ISASI04 _PPT.pdf

Wiegmann, D., Faaborg, T., Boquet, A., Detwieler, C., Holcomb, K. and Shappell, S. (2005). Human error and general aviation accidents: A comprehensive, fine grained analysis using HFACS. FAA Technical report no. DOT/FAA/AM-05/24. Available at: http://www.humanfactors.illinois.edu/Reports&PapersPDFs/Tech Report/05-08.pdf

Wild, D.A. (1997). Fratricide and the Operational Commander: An appraisal of losses to friendly fire. Technical report. Available at: http://www. stormingmedia.us/56/5615/A561523.html

Wilson, J.R. (2000). Fundamentals of Ergonomics in theory and practice. *Applied Ergonomics*, 31(6), 557–67.

Wilson, K.A., Salas, E. and Andrews, D. (2009). Introduction, in *Human Factors Issues in Combat Identification*, edited by D. Andrews, R.P. Hertz and M.B. Wolf. Aldershot: Ashgate.

Wilson, K.A., Salas, E., Priest, H.A. and Andrews, D. (2007). Errors in the heat of battle: Taking a closer look at shared cognition breakdowns through teamwork. *Human Factors*, 49(2), 243–56.

Woo, D.M. and Vicente, K.J. (2003). Sociotechnical systems, risk management, and public health: Comparing the North Battleford and Walkerton outbreaks. *Reliability Engineering and System Safety* 80, 253–69.

Woods, D.D. and Hollnagel, E. (2006). *Joint Cognitive Systems*. Boca Raton, FL: CRC Press.

Woods, D.D., Johannsen, L.J., Cook, R.J. and Sarter, N.B. (1994). *Behind Human Error: Cognitive Systems, Computers and Hindsight*. Ohio: CSERIC.

Yin, R.K. (2003). *Case Study Research: Design and Methods*. Third edition. Thousand Oaks, CA: Sage Publications.

Yule, S., Flin, R., Maran, N., Rowley, D., Youngson, G. and Paterson-Brown, S. (2008). Surgeons' non-technical skills in the operating room: Reliability testing of the NOTSS behavior rating system. *World Journal of Surgery*, 32, 548–56.

Yule, S., Flin, R., Paterson-Brown, S. and Maran, N. (2006). Non-technical skills for surgeons in the operating room: A review of the literature. Clinical Review. *Surgery*, 139(2), 140–49.

Zobarich, R., Bruyn-Martin, L. and Lamoureux, T. (2009). Team cognition during a simulated Close Air Support exercise: Results from a new behavioural rating instrument, in *Human Factors in Combat Identification*, edited by D.H. Andrews, P.H. Robert and M.B. Wolf. Aldershot: Ashgate.

Zobarich, R.M., Lamoureux, T.M. and Bruyn-Martin, L.E. (2007). Forward Air Controller: Task analysis and development of team training measures for Close Air Support. Defence R&D Canada – Toronto Contract Report. DRDC Toronto CR 2007–156. Available at: http://www.dtic.mil/cgi-bin/GetTRDoc?AD=ADA477167&Location=U2&doc=GetTRDoc.pdf

Zsambok, C.E. (1997). Naturalistic Decision Making research and improving team decision making, in *Naturalistic Decision Making*, edited by C.E. Zsambok and G.A. Klein. Mahwah, NJ: Erlbaum.

Index

Page numbers in *italics* refer to figures and tables.

For Product Safety Concerns and Information please contact our EU
representative GPSR@taylorandfrancis.com
Taylor & Francis Verlag GmbH, Kaufingerstraße 24, 80331 München, Germany